Environmental Management in Tropical Agriculture

Westview Replica Editions

The concept of Westview Replica Editions is a response to the continuing crisis in academic and informational publishing. Library budgets for books have been severely curtailed. Ever larger portions of general library budgets are being diverted from the purchase of books and used for data banks, computers, micromedia, and other methods of information retrieval. Interlibrary loan structures further reduce the edition sizes required to satisfy the needs of the scholarly community. Economic pressures on the university presses and the few private scholarly publishing companies have severely limited the capacity of the industry to properly serve the academic and research communities. As a result, many manuscripts dealing with important subjects, often representing the highest level of scholarship, are no longer economically viable publishing projects--or, if accepted for publication, are typically subject to lead times ranging from one to three years.

Westview Replica Editions are our practical solution to the problem. We accept a manuscript in camera-ready form, typed according to our specifications, and move it immediately into the production process. As always, the selection criteria include the importance of the subject, the work's contribution to scholarship, and its insight, originality of thought, and excellence of exposition. The responsibility for editing and proofreading lies with the author or sponsoring institution. We prepare chapter headings and display pages, file for copyright, and obtain Library of Congress Cataloging in Publication Data. A detailed manual contains simple instructions for preparing the final typescript, and our editorial staff is always available to answer questions.

The end result is a book printed on acid-free paper and bound in sturdy library-quality soft covers. We manufacture these books ourselves using equipment that does not require a lengthy make-ready process and that allows us to publish first editions of 300 to 600 copies and to reprint even smaller quantities as needed. Thus, we can produce Replica Editions quickly and can keep even very specialized books in print as long as there is a demand for them.

About the Book

Environmental Management in Tropical Agriculture
Robert J. A. Goodland, Catharine Watson, and George Ledec

Addressing the problem of the high cost of agricultural development in tropical regions, this book summarizes the environmental concerns associated with tropical agriculture. The authors highlight major environmental hazards confronted in tropical agriculture and suggest specific management options that could be used to reduce or avoid them. The first sixteen chapters systematically outline environmental aspects of a variety of tropical crops, including rice, maize, legumes, cassava, coffee, cocoa, sugar cane, tobacco, cotton, oil palm, rubber, timber, livestock, and fish. The following nine chapters focus on production factors and issues, such as soil erosion, efficiency of water and energy use, pest and weed management, and genetic diversity. Aimed at promoting the optimal and sustainable use of agricultural land through better environmental practices, this book offers guidelines for better ways to meet the world's growing demand for food and fiber in the context of improved environmental management. An underlying premise throughout the book is that prudent environmental and natural resource management is a prerequisite, not an impediment, to sustainable economic development.

This book is dedicated to the world's poor, with the aim that environmental management will help alleviate their plight. It is also dedicated to Mario Guimarães Ferri of São Paulo, Brazil for his work in neotropical and *cerrado* ecology, and to Otto Soemarwoto of Bandung, Indonesia for his in palaeotropical and *pekarangan* ecology.

Environmental Management in Tropical Agriculture

Robert J. A. Goodland,
Catharine Watson, and George Ledec

CRC Press
Taylor & Francis Group
Boca Raton London New York

CRC Press is an imprint of the
Taylor & Francis Group, an **informa** business

First published 1984 by Westview Press, Inc.

Published 2018 by CRC Press
Taylor & Francis Group
6000 Broken Sound Parkway NW, Suite 300
Boca Raton, FL 33487-2742

CRC Press is an imprint of the Taylor & Francis Group, an informa business

Copyright © 1984 Taylor & Francis Group LLC

No claim to original U.S. Government works

Visit the Taylor & Francis Web site at
http://www.taylorandfrancis.com

and the CRC Press Web site at
http://www.crcpress.com

Library of Congress Cataloging in Publication Data
Goodland, Robert J. A., 1939-
    Environmental management in tropical agriculture.
    (A Westview replica edition)
    Includes bibliographical references.
    1. Agricultural ecology--Tropics. 2. Agricultural development projects--
Environmental aspects--Tropics. 3. Agricultural pollution--Tropics.
I. Watson, Catharine. II. Ledec, George. III. Title.
S481.G66   1983   630'.913   83-14807

ISBN 13: 978-0-367-01546-6 (hbk)

ISBN 13: 978-0-367-16533-8 (pbk)

Contents

Figures

Acknowledgments

This book is a compilation, based entirely on the work of others. We want to acknowledge with gratitude all the generous help provided. Ralph Cummings, U.S. Agency for International Development; David Pimentel, Cornell University; Howard Irwin, Brooklyn Botanic Garden; Brian Trenbath, Imperial College; James Moomaw, International Agricultural Development Service; George Watson, tropical agriculture consultant; and Don King, U.S. State Department all reviewed the entire draft. However, the responsibility for any errors in fact or interpretation is our own. We are also grateful to our institutional colleagues for their support and guidance.

Robert J. A. Goodland
Catharine Watson
George Ledec

Abbreviations

BOD	Biochemical Oxygen Demand
BTU	British Thermal Unit(s)
Ca	Calcium
CIAT	Centro International de Agricultura Tropical, Cali, Colombia
CITES	Convention on International Trade in Endangered Species of Wild Fauna and Flora
cm	centimeter
FAO	United Nations Food and Agriculture Organization, Rome
g	gram(s)
ha	hectare(s)
HYV	High Yield Variety
ICAITI	Instituto Centroamericano de Investigacion y Tecnologia Industrial, Guatemala
IDRC	International Development Research Center, Ottawa
IRRI	International Rice Research Institute, Los Banos, Philippines
K	Potassium
kcal	kilocalorie(s)
kg	kilogram(s)
kJ	kilojoules
km	kilometer(s)
km^2	square kilometer(s)
lb(s)	pound(s)
l	liter(s)
m^2	square meter
m^3	cubic meter
Mg	Magnesium
Mn	Manganese
mg	milligram(s)
MMT	million metric tons
N	Nitrogen
P	Phosphorus
PANS	Pest and Agriculture News Service
pH	acidity/alkalinity scale
ppm	parts per million
RRIM	Rubber Research Institute of Malaysia
S	Sulfur
t	ton(s) (metric)
UN	United Nations
UNCTAD	United Nations Conference on Trade and Development, Geneva
UNDP	United Nations Development Program, New York
UNESCO	United Nations Educational, Scientific, and Cultural Organization, Paris
UNEP	United Nations Environment Programme, Nairobi
UNICEF	United Nations International Children's Emergency Fund
USDA	United States Department of Agriculture
WDM	World Development Movement, London
WHO	World Health Organization

Introduction

"It is right for lips and pens to move, provided
they also move hearts, purses, wills, funds, and
materials into action."

-- Andre van Dam,
Global Futures Network

Never before has our world had so much food available per
person and so many hungry people going without. Today, food is
often produced so expensively, or so far away, that the hungry
cannot afford it. This waste and suffering can be addressed in at
least two ways. The orthodox approach is to attempt to increase the
purchasing power of the poor. Widely followed, with mixed results,
this approach is not amplified here. This book suggests another,
complementary approach, favored by many environmentalists--reducing
the costs of agriculture in the tropics. Costs can be decreased by
reducing fossil fuel-based inputs such as biocides, inorganic
fertilizers, and diesel fuel, and by raising the efficiency of their
use; by recycling by-products (residues or "wastes"); by shortening
food chains; by promoting some measure of food and energy
self-reliance; and by minimizing damage to the renewable natural
resource base and the environment. Without attention to such cost
reduction, food demand by the additional two billion consumers
projected for the year 2000 may push food prices to even less
accessible heights, thus exacerbating hunger and even starvation.

Throughout the world in the future, conventional
agriculture will be forced by changing conditions to become more
natural resource efficient, or frugal and conserving. The factors
causing this change include the rising world population, with its
attendant strain on the assimilative capacity of the environment.
At the same time, the rising cost of energy has reduced the
attractiveness of using such high-energy, non-renewable agricultural
inputs as petroleum, fertilizers, and biocides.

While the Green Revolution has brought many dramatic
benefits in terms of rising productivity and increased yields, there
is now less promise that such yield increases will continue. As the
need for food grows ever more urgent, marginal land is increasingly
drawn into production; world yields per hectare are unlikely,

therefore, to rise as fast as they once did. Indeed, the rate of increase in per hectare yields of such basic crops as maize, wheat, and soybeans is declining. Even in the United States, agricultural productivity per unit of land area is levelling off. From an average annual rate of increase of 2.6 percent over the past 25 years, the growth in farm productivity per hectare slowed to 1.5 percent over the last 5 years (to 1981). More ominously, world per capita food output fell in 1981 from the three-year average, in spite of the overall decline in the world's rate of population growth. Even the annual growth in total grain production has declined by one third since 1973 (Brown, 1983).

In all agricultural areas, soil erosion and expanded human habitation also continue to reduce the productivity and availability of the land. Today the four principal biological foundations of the global economy--croplands, grasslands, forests, and fisheries--are under severe and increasing pressure. In many parts of the world, harvests exceed the regenerative and sustainable capacity of the environment, and in many cases the ecosystem is suffering irreversible degradation. It is essential for all those who are involved in agricultural development to acknowledge these trends and to promote solutions, some of which are suggested in this book.

In recent years, developing countries have shown increased awareness of the need to manage and sustain the renewable natural resources upon which their agricultural economies depend. According to the World Environment Center in New York, in 1982, 111 nations had environmental ministries or their equivalent, compared with only 11 in 1972. This increased acknowledgement of the problem suggests that there is still cause for hope.

Definitions: For the purposes of this book, we define **environment** to mean the aggregate of natural physical and biological conditions that are subject to human alteration. Environmental concerns encompass all living organisms, particularly humans, and especially include future generations. The environmental issue areas that broadly overlap with agriculture include human ecology (the relationship of human beings to the natural environment that sustains them); public and occupational health; pollution of air, water, and land; sustainable management of renewable natural resources; reduction of waste or improvement of natural resource use efficiency, through multiple use, recycling, and erosion control; conservation of unique natural ecosystems and their native plant and animal species; and aesthetic and cultural preservation.

By renewable natural resources we mean living resources (plants and animals) and other natural resources (particularly soil and water) that create or sustain life, and are self-renewing, if not overexploited or otherwise mismanaged. Nonrenewable natural resources are not self-renewing. They include minerals (which can often be profitably recycled) and fossil fuels (which cannot). Nonrenewable natural resources are the basis for most "modern" agricultural inputs. Care is needed in their use to prevent unnecessary environmental damage.

For the purposes of this book, we loosely define the tropics as areas with warm, year-round climates and rapid soil chemical processes. Virtually all the world's tropical areas are in developing countries.

Audience: We have written this book in response to increasingly urgent and numerous requests from colleagues in the tropics who are interested in or responsible for addressing the environmental implications of agricultural development. Ecologists and environmentalists concerned with the tropics should also find the book useful as an overview of environmental problems related to agriculture and the options available for addressing them. People in multilateral and bilateral development institutions, governmental environment and agriculture ministries, and private voluntary organizations should find the book useful in project design.

Objectives: This book discusses many of the significant and disturbing trends that have developed in tropical agriculture. We have three principal objectives. The first is to summarize the major environmental concerns related to tropical agriculture, which include various forms of environmental damage, as well as potential benefits. The second is to point out the options available for improving the environmental performance of tropical agriculture, by reducing the damage and enhancing the opportunities for benefits. Although these options are indicated throughout the book, their adoption depends on crucial economic, social, and political factors that generally are not discussed here. The third is to promote an increased general awareness of environmental concerns, particularly as they relate to sustainable economic development.

Each chapter presents choices, rather than rigidly prescribing policies. The chapters constitute a ready reference, compiled from many sources. On the assumption that most readers will be sufficiently familiar with tropical agronomy, discussion of purely agricultural concepts has been largely avoided. In the broadest sense, agriculture is management of the environment to facilitate the production of food and raw materials. Since the distinction between the agricultural and the environmental is indefinite (agriculture is sometimes referred to as applied ecology), occasional duplication and overlap are inevitable.

This book can be used at several levels. First, each chapter can stand on its own; chapters can be used individually as environmental checklists for the major tropical crops. Second, the book as a whole presents a general environmental overview of tropical agriculture. Our overall goals are to boost sustainable agricultural productivity, to improve agriculture projects by enhancing their potential benefits to local people, to prevent environmentally costly and detrimental side effects, and to reduce the waste of scarce natural resources.

This book deals with only some of many technical aspects which should be considered in any agricultural project design. Economic, political, and social factors, and details of soils or climate, for example, which vary in importance according to local circumstances, are scarcely discussed here. A short section on tobacco (Chapter 8) is the principal exception.

As with any list of precautions, the negative repercussions or the disadvantages of certain actions are emphasized, although the fundamental objective is to promote the optimal sustainable management of land and water resources. This becomes more important, yet more difficult, as agriculture extends increasingly into marginal lands. Ideally, where possible, crop choices and cropping systems should be adapted to the environment, rather than the conventional converse of modifying the environment to suit the crop.

Each chapter is deliberately succinct because we feel that the information conveyed will prompt recollection of experiences for many readers; for others, it may encourage consultation of the more detailed references provided in each chapter. Environmental management will be improved to the extent the proposed options are implemented. While incorporating these options into agricultural design is relatively straightforward, implementation will require close and systematic attention. Social, economic, and political processes are at least as important as specific environmental aspects and must be taken into account in agricultural decision-making, such as in determining tradeoffs between producing domestic food and export crops. Such processes, however, are beyond the scope of this book. Our thrust is that in order to meet the world's growing demand for food and fiber, agro-ecosystems must be environmentally sustainable.

References

Brown, L. R. 1981. Building a sustainable society. New York,
     Norton, 433 p.

Brown, L. R. 1983. Population policies for a new economic era.
     Washington, D.C., Worldwatch Institute Paper 53:45 p.

Clapham, W.B. 1980. Environmental problems, development and
     agricultural production systems. Environmental Conservation
     7(2):145-152.

Dumont, R. and Cohen, N. 1980. The growth of hunger. London, M.
     Boyars, 229 p.

Edens, T.C. and Koenig, H.E. 1980. Agroecosystem management in a
     resource limited world. BioScience 30(10):697-701.

USDA, 1981. World agriculture: outlook and situation.
     Washington D.C., USDA, Econ. Res. Serv. WAS-27:34 p.

1
Rice

"If you plan for one year, plant rice.
If you plan for ten, plant trees.
If you plan for one hundred years, educate
mankind."

--Kuan-Tzu

Although irrigated paddy rice (_Oryza sativa_) is one of the
world's most productive crops, its cultivation has relatively few
adverse effects on the environment. As a reflection of its benign
impact, rice has been grown continuously on the same land in some
regions (e.g., Luzon) with little decrease in yield for many
hundreds of years. Entire landscapes in Asia have been molded to
facilitate rice cultivation and to provide the precise water control
on which its growth depends. Environmental concerns associated with
rice production are restricted mainly to human diseases in
irrigation schemes and to increased biocide use. Upland and
rain-fed wetland rice production are not always so enviromentally
benign, however (see below). Moreover, additional land suitable for
the growing of rice of any kind is becoming increasingly scarce.
Future additional production from intensification or from increased
areas of upland rice, and from multicropping (in order to take
advantage of both wet and dry seasons) pose different risks. Such
augmentation will require improved natural resource management if it
is to be environmentally sustainable. Erosion, siltation, and
salinization, which can occur in rice cultivation, are discussed in
Chapter 20.

Rice provides the staple food for more than half the
world's population (more than any other single crop). As such, it
has been the subject of intensive research, centered at the
International Rice Research Institute (IRRI) in the Philippines,
aimed mainly at producing new high-yielding varieties (HYVs)
sometimes better named "high response varieties". Unfortunately,
high inputs of management technology, water, and agro-chemicals are
necessary for HYVs potential to be realized. These inputs have been
incorporated into "packages" of credit, extension services and
chemicals that accompany the HYV seeds. The HYVs, or their local
derivatives, are in use in most of the prime irrigated rice land and
have been most successful where farmers could afford the necessary

inputs. But now that the suitable land is mostly under cultivation, the spread of the HYVs has slowed and national production figures of rice-growing countries are levelling off. The impetus of the original breeding program is now being channelled into identifying genetic sources of disease and pest resistance, and of tolerance to low temperature, salinity, and soil toxicities.

The success of the Green Revolution in increasing rice production, and the convenience value of rice for home consumption (easy cooking and easy storage), have persuaded many non-Asian developing countries to extend their rice culture. Major areas for possible expansion lie in South America and West Africa. However, experience in research and extension in these new areas has uncovered unexpected and formidable difficulties (Buddenhagen & Persley, 1978). On the other hand, some Asian problem diseases (e.g., tungro, bacterial blight) fortunately, are absent, and some insects (e.g., planthoppers) have caused only minor damage so far.

Though some varieties of rice succeed in a wide range of humid environments, no varieties are tolerant of drought. Root systems in most varieties only exploit the top 20-25 cm of soil. Depending on the source and amount of water involved in rice culture, different water management problems arise, all of which have environmental implications.

Wetland Rice: Habitats suitable for this category include freshwater and marshes, swamps, mangrove swamps, seasonally-flooded plains or valley bottoms, and irrigated plains in semi-arid regions. Water management in each of these habitats presents its own difficulties. For example, in seasons of low river flow, salt water intrudes into mangrove soils and may lead to crop failure; if flood water rises faster than about 15 cm/day, the floating rice on a flood plain will be damaged. Hence, the variability of seasonal river flows may determine the advisability of using cleared mangrove swamps, and only where flash floods are absent is the production of floating rice advisable. Increased rice from converted mangrove swamps decreases shrimp and fish production offshore--a tradeoff to be addressed before such conversion (Chapter 12). Furthermore, people working in swamps and irrigation systems are exposed to the risk of water-related diseases such as schistosomiasis and malaria.

Wet-seeded rice needs accurately levelled land. This can be accomplished manually, assisted by draft animals, but when preparing large new schemes, the use of heavy earth-moving equipment is necessary. If the soil is not sufficiently clayey, puddling to reduce seepage losses will not be possible (African soils, for example, tend to be coarse-textured). Lateral seepage in soil or through bunds may make adjacent fields too wet for other desired crops. For example, when some farmers started growing a second (irrigated) rice crop in the Chiang Mai valley of Thailand, soybeans could no longer be grown by neighboring farmers. Excessive seepage will lead to loss of nutrients and possible eutrophication of water bodies.

Any irrigation system is prone to some degree of siltation. The accumulation of some silt in the field is beneficial in restoring depleted nutrients, though at the expense of upstream lands. Beyond a certain level of accumulation, however, the cost of clearing silted channels outweighs the advantage of the nutrient input itself. Watershed management is therefore an essential component of all irrigated rice projects (Chapter 20).

Hydromorphic or Rainfed Rice: This type of rice culture relies on the presence of flood waters during at least the early stage of growth. As the rainy season ends, however, the fields become progressively drier and the rice may suffer drought stress. Many African areas devoted to hydromorphic rice suffer from fluctuating water tables. Alternate drying and wetting of soils also leads to heavy losses of soil nitrogen and, as a result, efficiency of fertilizer use may be low in such soils.

Upland rice: While less important in terms of world output, upland rice is grown over large areas and often is the first crop to be raised after the forest or bush has been cleared. The environmental costs of forest loss are outlined in Chapter 17. In areas of high rainfall, or where forest is destroyed for only a few years of rice production (e.g., Amazonia) these deforestation costs are environmentally unacceptable. Like all rice varieties, upland rice is an annual crop which should be cultivated on open land free of weeds. However, this exposes the soil surface to the full impact of erosive rain during early stages of crop growth. Extensive cultivation of rice on land with gradients ranging up to 30° has led to greatly accelerated rates of erosion, with consequent losses of valuable nutrients and topsoil. After only a single crop under such conditions the soil may become exhausted, bedrock exposed, and the land left to be colonized by low value secondary scrub. In many parts of Asia, abandoned land is colonized by the aggressive, unpalatable weed grass, _Imperata_ _cylindrica,_ (lalang, or alang-alang) which both resists fire and is expensive to eradicate. Vast areas in West Africa enjoy a climate suitable for upland rice cultivation. However, because their soils are mostly coarse-textured and have low water-holding capacity, these areas are unlikely to produce sustained high yields of rice (Moormann & Veldkamp, 1978).

Soil Management and Nutrients: Wetland rice cultivation requires accurate land levelling, followed by preparation and maintenance of terraces and bunds to retain flood water. The extension of terracing up the sides of valleys is limited in large measure by the availability of soil. Where the soil is effectively puddled, weed control may be better and more water and nutrients will be retained. Seepage from rice fields can leach about 50 kg/ha of nutrients each year. Some of these lost nutrients may be replaced through silt input in the irrigation water and, if the level of water can be adequately controlled, useful quantities of nitrogen (up to 20 kg/ha/season) can be fixed by floating blue-green algae or _Azolla_ water ferns with their algal symbiont _Anabaena_ _azollae_. The great potential for biological nitrogen fixation in wetland rice systems has only recently been appreciated (Stewart, 1973; Lumpkin and Plucknett, 1980). Growth of the algae and _Azolla_ improves the nutrient cycle and any excess serves as valuable green manure.

If mangrove or other swamps are to be reclaimed for rice growing, it is vital first to check the concentration of sulphides in the soil. When sulphides (e.g., iron pyrites) present in an anaerobic soil ("cat clay") are oxidized after drainage, sulphuric acid is formed and soil pH falls to levels too low for rice, bringing aluminum and manganese into solution at toxic levels. Another soil toxicity problem is caused by high concentrations of ferrous iron in solution. In West Africa, for example, injurious concentrations are found in interflow water in areas characterized by senile soil landscapes and high rainfall. Flooding usually allows successful growth of wetland rice on acid soils by reducing any sulphate to sulphide and restoring pH to acceptable levels (Ponnamperuma, 1977).

Upland rice can be a component in highly diversified agro-ecosystems, such as exist in the Ivory Coast, where it is intercropped with yams and cassava, and in Northern Brazil, where it is intercropped with corn, beans, squash, bananas, and cassava. It also can be grown under perennial crops which cast only a slight shade, as in southern Thailand and Sumatra under young rubber trees and, in the Philippines, under coconut.

Since upland rice is an annual, cultivation on slopes exposes the land to erosion. Such practices as leaving rice straw and stubble after harvest, contour cultivation, and the construction of bunds and terraces will reduce erosion. In shifting cultivation, however, bunds or terraces are rarely built. If the soil and slope are such that erosion becomes serious, it is preferable to grow a different crop that will allow for better soil management, such as sweet potato planted on contour ridges.

Rice and Fish: Fishponds of _Tilapia_ and _Heterotis_ species can be maintained entirely on rice bran derived from polishing, so that with water-level control in wetland rice good enough to allow fish culture in the paddy fields, additional protein and vitamins can be obtained from the same tract. Rice fields can provide ideal habitats for fish rearing, which traditionally has been an important source of protein. All appropriate rice projects can provide for fish cultivation, retaining aquaculture traditions which avoid the use of chemicals or other practices likely to impede fish culture. Rice culture in Asia is associated with the rearing of domestic water fowl, usually ducks. Fowl droppings fertilize rice paddy water and promote the growth of nitrogen-fixing and other algae and bacteria, thereby augmenting rice production levels. Human sewage also can be used to advantage if appropriate health precautions are taken.

Pest Management: Insect pests are a major cause of yield losses in rice. An estimated 35 percent of potential yields in developing countries is lost yearly to pests of various sorts: insects and other arthropods, plant pathogens, rodents, and birds. Pre-harvest rice losses are estimated to be 35 percent for Asia, 21 percent for South America, 34 percent for Africa, and 21 percent for

North and Central America (FAO, 1978). Biocide use on rice is increasing, and though immediate yields may benefit, this may be accompanied by pest resurgence, development of insect resistance to biocides, destruction of economically useful insect predators, harm to aquatic animals in rice paddies, pollution, health hazards, and increased difficulty in controlling insect vectors of diseases such as malaria, dengue, and encephalitis (FAO, 1978). Biocide increase in rice cultivation is directly related to the development of high yielding rice varieties many of which lack resistance to major local insect pests and diseases. Moreover, the agronomic practices necessary for growing high yielding varieties often induce pest infestation. Heavily fertilized high yielding varieties typically have a high nutrient content, which promotes pest multiplication. These varieties of rice tend to be short-stemmed and to grow in a close canopy, creating an ideal microclimate for many pests and diseases. Pest control is failing in some areas because the prophylactic use of biocides has led to resistance. Multiple cropping of rice has also significantly increased pest populations.

FAO's Guidelines on Integrated Control of Rice Insect Pests (1978) lists the following examples of problems that have developed as a result of reliance on biocides in rice production:

Pest Resurgence: Heavy use of biocides in Japan against rice borers led to their resurgence, as well as resurgence of leaf and plant hoppers. Concomitant outbreaks and changes in the status of minor pests were largely due to the destruction by biocides of natural enemies, particularly of the egg parasites of borers and the spider predators of hoppers. Regional aerial biocide applications against borers in Indonesia destroyed some natural parasites and led to outbreaks of the rice gall midge.

Insect Resistance: The most important cases of biocide-induced pest resistance have occurred in the green rice leaf hopper (Nephotettix species) and the brown plant hopper (Nilaparvata lugens) in Japan and Indonesia. One area of Japan experienced a tenfold increase in the resistance of Nephotettix to malathion. Resistance has also been reported in the rice stem borer (Chilo suppressalis), the rice leaf miner (Agromyza oryzae), and the rice leaf beetle (Ouelma oryzae) in Japan (Chapters 17 and 18).

Traditional methods of cultivating local low-yielding varieties are characterized by minimal chemical use, a considerable degree of natural control of rice pests, and some resistance of these cultivars to infestations. Natural enemies of rice pests are abundant in the rice fields of Asia. Before biocides were introduced into Japan, cultural practices such as regulation of planting time effectively controlled rice pests, particularly stem borers, leaf miners, and stem maggots. FAO (1978) reports that effective rice protection programs based on resistant varieties and various cultural practices have been undertaken in the Philippines, India, Egypt, Latin

America, and the United States. Integrated pest control for most of the rice-producing areas of China is based on monitoring, pest forecasting, cultural control practices, and limited use of chemicals, in addition to biological control agents.

Weed Management: Most of the labor involved in traditional rice culture is applied to weeding (250-780 hours per person of hand weeding/ha/crop). Densities as high as 10,000 weed stalks per square meter can occur. If weeding is neglected, yield is correspondingly reduced--and when the crop is badly infested with wild rice, the yield loss may even be total.

The frequency of weeding (hand or chemical) depends on the type of rice culture: rainfed rice may need three or more weedings, while upland rice may need only two; under flood irrigation conditions, a 10 cm depth of water controls many weeds. The type of weed species depends on the water regime of the field. Thus, in upland cultivation, Rottboellia exaltata and various species of Commelina, Digitaria and Echinochloa may predominate. These weeds are resistant to many common herbicides and hand-pulling may be necessary. If perennial weeds like Cyperus rotundus and Imperata cylindrica predominate, the land may have to lie fallow for several years; mechanical cultivation when preparing a seedbed can disseminate propagules and thus greatly multiply the number of Cyperus rotundus plants. In rainfed and irrigated paddy cultivation, these species are joined by water-loving species, particularly the wild rices (as at Jari in Brazil). Herbicide action does not persist well under moist, warm conditions, and no herbicide is sufficiently selective to distinguish between wild and cultivated rice.

In addition to hand or machine hoeing, non-chemical weed control methods include controlled flooding, high crop-planting density, planting in rows to facilitate later hoeing, rotating wetland rice with an upland crop (e.g., a legume), use of an intermediate-height high-tillering variety, banding split dressings of fertilizer next to the crop plants rather than by broadcasting, and wet seeding or transplanting in order to give the crop plants a head start.

In addition to the financial expense of biocides, their use often imposes environmental costs such as killing non-target organisms and damaging human health (Chapter 17). Herbicides may also increase crop susceptibility to insect and disease pests. Rice stemborer damage, for example, increased as much as 45 percent when the herbidice 2,4-D was used on rice. Maize becomes more susceptible to insects and diseases when exposed to normal dosages of 2,4-D. Thus, the trade-off between biocide use and increased susceptibility of crops to pests should be carefully weighed before deciding on the appropriate mix of weed control methods.

References

Akubundu, I.O. and Fagade, S.O. 1978. Weed problems of African rice lands. (29-43), in Buddenhagen, I.W. and Persley, G.J. (eds.) Rice in Africa. London, Academic Press, 356 p.

Anon, 1978. Proceedings of the workshop on the genetic conservation of rice. Los Banos, Laguna, Phillipines, IRRI:60 p.

Argullo, E., De Padua, D.B. and Graham, M. (eds.) 1976. Rice: postharvest technology. Ottawa, International Development Research Centre, 394 p.

Asian Productivity Organization, 1977. Farm water management for rice cultivation. Tokyo, Asian Productivity Organization, 333 p.

Bhatty, I.Z. 1976. The impact of the price rise in petroleum based agricultural inputs on the production of wheat and rice in India. London, Commonwealth Secretariat Economic Paper 6:74 p.

Brown, L.R. 1970. Seeds of change: the green revolution and development in the 1970's. New York, Praeger, 205 p.

Buddenhagen, I.W. and Persley, G. J. (eds.) 1978. Rice in Africa. London, Academic Press, 356 p.

Buddenhagen, I.W. 1978. Rice ecosystems in Africa in Rice in Africa. London (see above).

Chandler, R.F. 1979. Rice in the tropics: a guide to the development of national programs. Boulder, Colorado, Westview Press, 256 p.

Chiang, H.S., Ryv, J.H., Jo, J.S., and Kwon, T.W. 1978. Nutritional losses during washing and cooking of rice. Extension Bulletin (ASPAC/FFTC) 103: 9 p.

Colombo, V., Johnson, D.G. and Shishido, T. 1978. Reducing malnutrition in developing countries: increasing rice production in South and South-east Asia. New York, Trilateral Commission, 55 p.

Dalrymple, D.G. 1975. Measuring the green revolution: the impact of research on wheat and rice production. Washington, D.C., USDA/USAID:40 p.

De Datta, S.K. de, Bolton, F.R., and Lin, W.L. 1977. Prospects for using minimum tillage and zero tillage in tropical lowland rice. Weed Research (UK) 19(1):9-15.

FAO, 1978. Guidelines for integrated control of rice insect pests. FAO, Rome, 115 p.

Farmer, B.H. (ed.) 1977. Green revolution? Technology and change in rice growing areas of Tamil Nadu and Sri Lanka. Boulder, Colorado, Westview Press, 429 p.

Framji, K.K. (ed.) 1977. Irrigated rice: a world-wide survey. New Delhi, International Commission Irrigation Drainage, 705 p.

Freedman, S.M. 1980. Modifications of traditional rice production practices in the developing world: an energy efficiency analysis. Agro-Ecosystems 6(2):129-147.

Grist, D.H. and Lever, R.J.A.W.1969. Rice. London, Longmans, 520 p.

Hanks, L.M. 1972. Rice and man: agricultural ecology in South-east Asia. Chicago, Aldine-Atherton, 174 p.

International Rice Research Institute (IRRI), 1975. Changes in rice farming in selected areas of Asia. Los Banos, Philippines, IRRI, 377 p.

IRRI, 1976. Symposium on climate and rice. Los Banos, Philippines, IRRI, 565 p.

IRRI, 1979. Research highlights for 1978. Los Banos, Philippines, IRRI, 118 p.

IRRI, 1981. Energy requirements for alternative rice production systems in the tropics. Los Banos, IRRI Res. Paper 59:14 p.

Khan, M.H. 1975. The economics of the green revolution in Pakistan. New York, Praeger, 229 p.

Kiritani, K. 1977. Recent progress in the pest management for rice in Japan. JARQ (Japan) 11(1): 40-49.

Lumpkin, T.A. and Plucknett, D.L. 1980. Azolla: botany, physiology, and use as a green manure. Economic Botany 34(2):111-153.

Lumpkin, T.A. and Plucknett, D. L. 1983. Azolla: as a green manure. Boulder, Colorado, Westview Press, 230 p.

Litsinger, J.A. and Moody, K. 1976. Integrated pest management in multiple cropping systems. (293-316) in Papendick, R.I., Sanchez, P.A. and Triplett, G.B., (eds.) Multiple Cropping. Madison, Wisconsin, Amer. Soc. Agron: 378 p.

Mears, L.A. and Marquez, R.C. 1972. Rice storage and milling. Quezon City, University of the Philippines, Institute of Economic Development and Research, Discussion Paper 72-11:52 p.

Mitsuda, H. 1965. Enrichment of rice by soaking method (521-528) in Proc. Inst. Int. Congr. of Food Sci. and Technol. New York, Gordon and Breach, 2 vols.

Moorman, F.R., Veldkamp, W.J. and Ballaux, J.C. 1977. The growth of rice on a toposequence. Plant and Soil 48(3):565-580.

Moorman, F.R. and Veldkamp, W.J. 1978. Land and rice in Africa: constraints and potentials. (181-192) in Buddenhagen, J.W. and Persley, G.J., (eds.) Rice in Africa. London, Academic Press, 356 p.

Palmer, I. 1976. The new rice in Asia: conclusions from four country studies. Geneva, United Nations Research Institute for Social Development, Report 76(6):146 p.

Pathak, M.D. and Dyck, V.A. 1973. Developing an integrated method of rice insect pest control. Pest and Agric. News Service 19(4):534-544.

Ponnamperuma, F.N. 1977. Physico-chemical properties of submerged soils in relation to fertility. IRRI Research paper series 5:32 p.

Stewart, W.D.P. 1973. Nitrogen fixation by photosynthetic micro-organisms. Ann. Rev. Microbiol. 27:283-316.

Talley, S.N. et al. 1977. Nitrogen fixation by Azolla in rice fields. (259-281) in Hollaender, A. (ed.) Genetic Engineering for Nitrogen Fixation. New York, Plenum, 538 p.

Wang, J.K. and Hagan, E.R. 1981. Irrigated rice production systems design procedures. Boulder, Colorado. Westview Press, 300 p.

Yoshida, 1981. Fundamentals of rice crop science. Los Banos, IRRI: 269 p.

2
Maize, Sorghum, and Millet

Maize (<u>Zea</u> <u>mays</u>) and sorghum (<u>Sorghum</u> <u>vulgare</u>) rank third and fourth in world production among major cereals, after wheat and rice. Millet (<u>Setaria</u> <u>italica</u>) is an important food source in India, Afghanistan, Iran, and Turkey. It is also a major crop in the Sahelo-Sudan region, where it is grown in areas receiving too little rainfall to support other crops. Sorghum is also well-adapted for cultivation in semi-arid regions. Though relatively water-demanding, maize excels as a grain crop in drier regions of the United States, where it is grown under irrigation. However, in developing countries, maize is rarely irrigated. It is usually grown intensively, being relay-cropped in Latin America with beans, and in Africa with cassava, plantains, and other staples. As maize, sorghum, and millet are staple crops in vast areas of the developing world, and particularly as sorghum and millet are uniquely adapted to dry regions where food shortages are chronic, improved production of these crops can decrease hunger and malnutrition.

Soil Management: Soil erosion and exhaustion are inherent in cultivating these three crops, as with all intensively-grown annual row crops. Serious environmental problems ensue when cultivation of these crops is extended into excessively arid areas, displacing traditional grazing. Consequent overgrazing on the remaining grassland areas damages pasture and browse shrubs and exacerbates desertification, as already has occurred in the Sahel.

Where population pressures and inequitable land tenure systems compel families to cultivate steep slopes, the traditional maize crop is associated with considerable soil loss. Heavy, sustained rainfall causes much of the erosion and loss of nutrients in wetter areas, erosion due to sporadic but intense rains and wind is characteristic of semi-arid, coarse grain-growing regions. African regions that produce sorghum and millet frequently are the areas most threatened by desert encroachment and extreme land degradation.

Maize stalks and leaves are used as cattle feed, fuel, thatch, and fencing in both Latin America and Africa. Plowing under maize field residue or fertilizing with ash do not seem to be common practices. One report states that in Africa ...

"debris left on fields from previous crops is burned to get ash for fertilizing fields, but often by the time fields must be prepared all maize stalks from the previous crop have already been used for fuel and little debris is left. Goat manure is applied to fields near the homestead, but typically is not carried to distant fields or to those on slopes above the homestead, because beasts of burden are not available" (Miracle, 1966).

Maize, sorghum and millet respond dramatically to fertilizer when rainfall is adequate, and, in general, the lower the inherent fertility of the soil, the greater will be the response. Arnon (1972) reports that ...

"Trials (of millet) in the summer-rainfall region south of the Sahara have shown that the application of nitrogen alone increased yields from 100 kg/ha (normally obtained on unfertilized soils) to 1,500 kg/ha. ... NPK, applied under favorable growing conditions, makes possible yields of approximately 2,000 kg/ha, when there is 175 mm of available soil moisture for the growing season."

While massive amounts of fertilizer are used to grow grain in the developed world, even for cattle feed, virtually no fertilizer is available, for economic reasons, to increase production of sorghum, millet and maize in the developing world to feed directly millions of people. Traditionally, farmers unable to afford fertilizer have had to turn to relay-planting and intercropping of legumes to recharge the soil nitrogen necessary to sustain their low yields. Thus, in Colombia, 98 percent of the bean crop is grown association with maize; in tropical Latin America as a whole, 60 percent of the maize is associated with other crops, usually legumes (Francis, et al., 1976). Agboola and Fayemi (1971) report that the yields of four successive maize plantings intercropped with a legume but without nitrogenous fertilizer were comparable with yields from maize grown with the nitrogenous fertilizer. Experiments in Africa, India, Thailand, and the Philippines show that, especially on low nitrogen soils, yields from millet-legume and sorghum-legume mixtures are regularly 20 to 60 percent greater than those obtained from comparable pure stands (Chapter 3).

Pest and Weed Control: Until recently, comparatively little biocide was applied to maize and even less to sorghum, mainly because of the economics of production. Present use is normally confined to one dusting with aldrin at sowing. Alternative methods have been used to control several key pests and diseases of these crops, including use of resistant plant varieties, crop rotation, selection of tillage method and time of planting, and encouragement of natural enemies. However, in some tropical countries, maize now receives heavy preventive treatments with biocides. Furthermore, biocides are being used in increasing quantities on both maize and sorghum in tropical areas, particularly Latin America. The need to prevent such over-dependence on biocides is paramount (Chapter 17). Only selectively-acting biocides, as part of a system of integrated pest management (IPM), can be recommended (Chapter 18). The global area under maize, sorghum, and millet cultivation is so vast when compared, for example, with the area under biocide-intensive cotton cultivation, that the amounts of chemicals involved in even moderate biocide applications on these crops will become enormous if these trends continue.

Maize, sorghum, and millet have many pests in common. Areas devoted to the production of each crop often overlap, extending from the wettest zones where maize predominates, to the driest where millet is frequently the sole cereal grown. The three crops are often planted together in mixed or multiple cropping systems, and this diversity decreases pest incidence as well as the risk of overall crop failure. The overlapping pest complexes and agro-ecological conditions support the inclusion of sorghum and millet along with maize, in combined pest management research programs, although in the most arid regions research will remain closely crop-specific. Where chemical controls are unavoidable, all precautions should be taken to ensure human safety and health and to avert environmental contamination (Chapters 17 and 18).

Certain birds, especially the _Quelea_ species in Africa, take a heavy toll on ripening sorghum and millet. Control of breeding habitat may be the most effective way to minimize this damage. Rats and other rodents are often also a serious problem. Improved rodent-proof granaries are a partial solution. Harmless snakes encouraged to live in grain-growing areas will reduce grain losses due to rodents. Enhancing the nesting habitat for owls and other birds of prey is also useful in this regard. The slaughter of rodent-hunting snakes and birds of prey should be strongly discouraged.

Although herbicides are used extensively in maize and sorghum cultivation throughout the developed world, their use is rare in tropical countries. However, recent trends in no-till agriculture, where needs are controlled chemically instead of by cultivation, poses the prospect of more extensive herbicide usage.

The Environmental Implications of Genetic Breeding Programs: Genetic breeding programs have led to vast increases in maize yields in the developed world. Similar programs have recently been started for sorghum and millet. The development of early-maturing and insect-resistant cultivars will most likely be the major contribution toward increasing sorghum and millet production.

The breeding programs of the Green Revolution have been a mixed blessing. Although capable of high yields, many high-response varieties reach maximum productivity only under intensified cultivation with high levels of inputs. Some improved strains are less drought- and pest-resistant than the original varieties; shorter-strawed varieties provide less cooking fuel and fodder and are susceptible to weed invasion. Furthermore, the substitution of improved strains for traditional crop varieties has inadvertently led to a loss of germplasm--a risky narrowing of the gene pool of the traditional varieties on which all research for new, improved strains must be based (Chapter 24).

Thus, while genetic breeding toward improved strains should be a high research priority, precautions should be taken to ensure that germplasm is not irrevocably lost. Many national research projects have begun intensive collection and preservation programs. The international agricultural research centers under the Consultative Group on International Agricultural Research (CGIAR), particularly the International Board for Plant Genetic Resources, have become involved with germplasm preservation in several tropical countries.

High-response varieties requiring substantial fertilizer and (irrigation moisture supplements) continue to be developed. However, other less demanding varieties also merit research priority, since they are suited to the harsh conditions under which most subsistence agriculture occurs. We often forget that in most of the world, reliability and stability of crop yields is more important than record production.

References

Agboola, A.A. and Fayemi, A.A. 1971. Preliminary trials on the intercropping maize with different tropical legumes in Western Nigeria. J. Agric. Sci. Camb.77:219-225.

Areekul, S. 1965. Insect pests of corn in Thailand. Bangkok, Dept. of Entomology and Plant Pathology, Kasetsart University, 204 p.

Arnon, I. 1972. Crop production in dry regions. London, Leonard Hill, 2 vols.

Dendy, D.A. (ed.) 1977. Proceedings of a symposium on sorghum and millets for human food. London, Tropical Products Inst:138 p.

Doggett, H. 1970. Sorghum. London, Longmans, 403 p.

Etasse, C. 1977. Sorghum and pearl millet. (27-39) in Leakey, C. L. A. and Wills, J.B. (eds). Food Crops of the Lowland Tropics. London, Oxford University Press, 345 p.

FAO Panel of Experts on Integrated Pest Control. 1975. The development and application of integrated pest control in agriculture. Rome, FAO:39 p.

Francis, C.A., Flor, C.A., and Temple, S.R. 1976. Adapting varieties for intercropping systems in the tropics (235-253) in Papendick, P.A. et al. (eds.). Multiple Cropping. Madison, Wisconsin, Amer. Soc. Agron: 378 p.

Kerr, A.D. 1979. Food or famine: an account of the crop science program supported by International Development Research Center. Ottawa, 79 p.

Knopacheer, H. and Menz, K. 1980. Benchmark surveys of three crops in Nigeria: wheat, millet, sorghum. Ibadan, Nigeria, 28 p.

Miracle, M. P. 1966. Maize in tropical Africa. Madison, University of Wisconsin Press, 327 p.

Pava, H.M. 1978. Grain yield per day of sorghum in the tropics. Laguna, Kalikasan, 7(3):259-268.

Payak, M.M. 1975. Research on diseases of maize: coordinated maize improvement scheme. New Delhi, Indian Council of Agricultural Research, 228 p.

Philippine Council for Agricultural Research. 1972. Corn and
    sorghum: national program of research. Los Banos, Univ. of the
    Philippines, 30 p.

Salazar, R. 1971. El maiz: la planta mas humana. Mexico D.F.
    Libreria de M. Porrua, 187 p.

3
Legumes, Oilseeds, and Vegetables

Legumes: Legumes, which include major oil producing crops and several vegetables, are strongly recommended for development projects on environmental grounds. Their capacity to fix nitrogen is valuable in tropical soils. Crop rotations that include legumes enrich the soil with 15-40 kg of organic nitrogen per hectare. The tropical grain legumes are major sources of protein and include the two most important oilseeds, soybeans (**Glycine max**) and peanuts (**Arachis hypogaea**). Peanuts comprise about 60 percent of all tropical grain legumes and are produced in semi-arid regions, mostly as a cash crop for export. Peanuts are also a major source of domestic food and cooking oil (Rachie and Roberts, 1974).

Other important legume species (e.g., cowpea **Vigna unguiculata**, pigeon pea, mung bean, and lima bean) supply protein and are major secondary crops in both Asia and Africa. **Phaseolus** beans are commonly grown in Central and South America in mixed stands with maize and are a principal source of local protein. Most widely-grown cover crops are legumes which furnish vegetative ground cover and serve to control erosion, but do not supply food (Chapters 1 and 25). In addition to controlling erosion, other legumes produce natural gums, fiber, starch, fruits, timber and green manure (NAS, 1979). A variety of non-leguminous tropical crops remain underexploited for starch, fruits, vegetables, oilseeds, forage and other uses (NAS, 1975).

The tropical lowland leguminous shrub, Ipil-Ipil, (**Leucaena leucocephala**) has great potential as animal feed, fuelwood, forage and human food, while also controlling erosion and improving soil fertility in degraded tropical lands (NAS, 1975; NAS, 1977). In a single year's growth, **Leucaena** yields 94 kg N/ha, as well as up to 2,240 kg/ha of leaves (edible by livestock) plus 13,330 poles of 3 cm diameter yam stakes in Nigeria (IITA, 1981).

The winged bean (**Psophocarpus tetragonolobus)** has recently received increased attention for its high protein yields per hectare from its leaves, pods, seeds, and fleshy root (PCARR, 1978). Grown as a backyard garden crop, the winged bean provides a low-cost, easily accessible source of high protein foods.

Leguminous crop residues furnish valuable fodder for animals. They can also be added to the soil to enhance organic N content. Even when this is done however, yields are usually low unless and supplemental mineral fertilizer is added. Phosphates are normally the only fertilizer consistently recommended for legume crops.

Oilseeds: 1980 world production of oilseeds exceeded 170 million metric tons. In recent years, large increases in production have come from soybeans in Brazil, oil palm in Asia and West Africa, and from sunflower (<u>Helianthus annuus</u>) in the temperate zones. Sesame (<u>Sesamum indicum</u>) and peanut oil supplies are stable and world demand for coconut has declined. World demand for most oilseeds continues to rise faster than population, but food uses for vegetable oils are no longer predominant. Current interest in finding petroleum substitutes will probably strengthen vegetable oil demand through substitution, but other latex and oil crops may eventually prove superior.

The jojoba shrub (<u>Simmondsia chinensis</u>), native to the Sonoran Desert of northwestern Mexico and the southwestern United States, and cultivated since 1977 in the Sudan, promises to be the first zero-calorie cooking oil (being a liquid wax with no glycerol), as well as providing a very high quality industrial oil and protein-rich residues. Jojoba thrives in poor soil with as little as three or four inches of rain per year, in part due to its phreatophytic root, which can be 100 feet long. One thousand jojoba bushes per acre produce 3,000 lbs. of seeds per year (averaging 53 percent oil content). At the 1982 price of US$20/lb, this would yield US$200 per gallon or US$8,000 per barrel.

Vegetables: Vegetable crops are frequently grown where irrigation is developed in proximity to major markets. If properly located in combination with transportation, marketing, and extension services, these crops can contribute substantially to small farmer incomes. Net returns to small farmers may exceed US$5,000/ha/yr (Villareal, 1980).

Vegetable crops are nutritionally advantageous to consumers, since they provide humanity's major source of vitamins A and C, many minerals, and much protein. A diverse list of about sixty crop plants makes up the vegetable group but only about fifteen are widespread in the tropics. Edible green leaves of tropical plants are a significant food source but are generally not widely used (Martin and Ruberte, 1979).

Producing vegetable crops is one of the most demanding of agricultural enterprises and requires highly developed skills and good quality land. In most cases, these crops also require high levels of inputs such as 800-1200 kg N/ha/year and multiple applications of biocides for control of major pests and diseases. They may contribute to soil and water pollution at a greater rate than cotton production, but frequently are of higher economic value and can readily support the costs of chemical inputs. Five or six crops may be harvested annually from the same area (Menegay, 1976).

Many countries now regulate the quantities, types and timing of biocide applications to vegetable crops before they enter the market, in order to protect consumers from toxic materials which do not degrade. Increasing demand is being made by farmers and regulatory agencies for biocides that are biodegradable and relatively less toxic to humans. Integrated pest management (Chapter 18) practices are now being developed which will permit safer and more environmentally sound pest control practices without sacrificing yields. For example, Cowpea podborer (Maruca) resistance has been identified and bred into new cowpea varieties.

References

Allen, O.N. and Allen, E.K. 1981. The Leguminosae: a source book of characteristics, uses and nodulation. Madison, University of Wisconsin, 812 p.

Asian and Pacific Council (ASPAC). 1974. Multiple cropping systems in Taiwan. Taipei, Food and Fertilizer Technology Center, 77 p.

AVRDC, 1978. Progress Report '77. Asian Vegetable Research and Development Center. Shanhua, Taiwan, 90 p.

Dalrymple, D. 1971. Survey of multiple cropping in less developed nations. Washington, D.C., Foreign Economic Development Service, U.S. Department of Agriculture/U.S. Agency for International Development:108 p.

Duke, J.A. 1981. Handbook of legumes of world economic importance. New York, Plenum, 345 p.

Gomez, A.A. and Zandstra, H.G. 1977. An analysis of the role of legumes in multiple cropping systems for small farmers. Honolulu, Univ. Cooperative Extension Serv. 145:81-95.

Grubben, G.J.H. 1975. The culture of the amaranth, tropical vegetable leaves. Wageningen, Venman and Zonen, 223 p.

IITA, 1981. Research highlights for 1980. Ibadan, Nigeria: International Institute of Tropical Agriculture:64 p.

Martin, F.W. and Ruberte, R.M. 1979. Edible leaves of the tropics. Washington, D.C., Mayaguez Inst. Trop. Agric. 234 p.

Menagay, M. R. 1976. Farm management research on cropping systems. Shanhua, Taiwan: Asian Vegetable Research and Development Center Technical Bulletin 2:19 p.

NAS, 1975. Underexploited tropical plants with promising economic value. Washington, D.C., National Academy of Sciences, 189 p.

NAS, 1977. Leucaena: promising forage and tree crop for the tropics. Washington, D.C., National Academy of Sciences, 115 p.

NAS, 1979. Tropical legumes: resources for the future. Washington, D.C., National Academy of Sciences, 331 p.

Nath, P. 1976. Vegetables for the tropical region. New Delhi, Indian Counc. Agric. Research, 109 p.

Oomen, H.A.P.C. and Grubben, G.J.H. 1978. Tropical leaf vegetables in human nutrition. Amsterdam, Royal Tropical Inst., 140 p.

PCARR, 1978. The winged bean: the first international symposium on developing the potentials of the winged bean. Manila, Philippines: Philippine Council for Agriculture and Resources Research, 447 pp.

Price, E.C. 1977. Multiple cropping in tropical Asia. in Stelly, M. (ed.) Multiple Cropping. Amer. Soc. Agron. 378 p.

Rachie, K.O., and Roberts L.M., 1974. Grain legumes of the lowland tropics. Advances in agronomy 26: 132 p.

Villareal, R. L. 1980. Tomatoes in the tropics. Boulder, Colorado, Westview Press, 174 p.

Vegetables of the Tropics, 1980. Tokyo, Norinsho Netta: Nogyo Kenkyu Senta, 716 p.

Weiss, E.A. 1983. Oilseed crops. New York, Longmans, 608 p.

4
Cassava and Other Root Crops

Cassava or manihot (**Manihot esculenta**), a major tropical root crop, ranks among the world's major staple foods, with the world harvest reaching 100 million metric tons in 1976. Its many adaptive features enable it to withstand drought, brief periods of flooding, generally low soil fertility, and light management. In terms of calories produced per unit area per unit time, cassava is one of the most productive of all crops. Yields of 10-20 t/ha per harvest are commonly reported under field conditions, whereas yields of 77 t/ha per harvest or about 211 kg/ha/day are possible under experimental conditions. This is more than one-third greater than calorie production from rice grown under optimal conditions.

Cassava, normally a rain-fed crop, is easily planted and requires few inputs and little manual attention (besides weeding). Because cassava preserves well for long periods while still in the ground, it is used as a food source during periods of low production. Their tubers and deep phreatophytic roots enable cassava and other root crops to tolerate periods of drought. Latex contained in the stems and leaves possibly increases drought tolerance, to the extent that it reduces transpiration. Some manihot species produce latex sufficient to be commercial sources of rubber (e.g., **Manihot glaziovii** yields "Ceara" rubber).

Other root crops, including yams, cocoyams and sweet potatoes, also are well adapted to humid tropical conditions. Highly organized societies based on vegetatively propagated root crops have evolved in many parts of the forest zone of the humid lowland tropics, particularly in areas of West Africa, Melanesia, and Polynesia. Although providing valuable alternative sources of carbohydrates, most species of potatoes require more care and labor than cassava. In general, potatoes are more prone to pest attack than cassava, although sweet potato has notably fewer yield-diminishing diseases than other potatoes.

Nutritional Value of Cassava: The main traditional product, the root tuber, is largely digestible starch, with 0.5-1.5 percent protein; sugar (5 percent) and fiber (1-2 percent) comprise the remainder, as with most such tubers. Dietary lack of protein (e.g., soybean protein, Grace 1977), and over-reliance on root crops can lead to severe malnutrition, including kwashiorkor in children. In iodine-deficient areas, the cyanogenetic glycoside linamarin that is present in the cassava root can inhibit thyroid activity (IDRC, 1981). The great advantage of exceptionally high yields of calories from cassava, therefore, is tempered by low toxicity and its low protein content. However, if the starchy root is mixed with yeasts and ptyalin (as in human saliva), the partially digested gruel product contains improved levels of vitamins and amino acids. Many traditional societies take advantage of this and consume much of their tuber intake as fermented gruel (e.g., gari in Nigeria; cassiri in Guyana). Furthermore, cassava foliage, often discarded in some countries, has very high protein levels so that the leaves may be eaten, or the protein extracted industrially. Even stems contain up to 10 percent protein in total dry weight. A factory in Thailand, for example, uses cassava foliage for the manufacture of protein pellets for export. Similar projects are being undertaken in Colombia and Venezuela. Because of its great adaptability, cassava can be grown in association with other more nutritious food crops, such as pulses.

Cassava is widely used in most tropical areas in animal feed. Cassava pellets or chips can be substituted for cereals to comprise up to 70 percent of livestock diets. However, animals consuming excessive proportions of unprocessed cassava suffer from a subclinical toxicity from residual hydrocyanic acid.

Nutritional Value of Other Root Crops: Because cassava is the dominant root crop in many tropical regions, root crops generally are often erroneously associated with low protein yields. Actually, yams, sweet potatoes and Irish potatoes may yield significantly more protein per unit of land than do maize, rice, or sorghum, while cocoyams yield an approximately equivalent amount (Table 3.1). The tubers of these other root crops may contain 5%

protein which, in the case of sweet potatoes, contains most of the essential amino acids. Yellow-fleshed varieties of sweet potato provide abundant carotene (vitamin A). The leaves of sweet potato contain very high protein levels (27 percent), and are eaten by humans as well as by livestock. Yams, sweet and Irish potatoes, and cocoyams yield about half as many calories per unit area of land as do grains and as much as eight times as many calories per unit land as do soybeans.

Figure 3.1: Calorie and Protein Productivity of Various Food Crops in West Africa

	Yield (kg/ha)	Calories (millions/ha)	Protein Production (kg/ha)
Cassava	6813	8.2	37
Yam	-	5.7	107
Sweet potato	6011	7.4	96
Irish potato	8181	4.7	128
Cocoyam (taro,tannia)	-	4.5	80
Maize	1203	3.2	82
Rice	1704	3.2	72
Sorghum	704	2.4	70
Soy beans	479	0.8	78

Sources: Leakey and Wills (1977); FAO 1977 Production Figures for Africa.

Soil Management: The principal environmental concerns associated with root crops are erosion (particularly on sloping land) and soil depletion caused by the churning and exposing of soil during and after harvest of the deep roots. During early growth, leaf cover is incomplete, again exposing soil to erosion unless precautions are taken. These can include: strip-cropping, using legumes and vines as alternate crops; row planting along contours, using mulches and "no-till" methods; intercropping with perennials (often under the canopies of tree crops such as coconuts); and using cover or relay crops. Deforestation caused by shifting cultivation or by expansion of permanent cultivation can have the worst environmental impact (Chapter 21).

A comparative study of corn, sugar cane, cassava, bananas, and cabbage has shown that cassava is not the most nutrient-depleting of these crops (Hongsapan, 1962). Cassava offers reasonable yields on soil that is too deficient in nutrients to sustain other crops. In traditional shifting cultivation systems, cassava is often the last crop to be cultivated before fertility declines to uneconomic levels and land is left to bush regeneration. When the root is converted into oligotrophic, storable commodities such as starch, flour, bread or alcohol, most mineral nutrients can be recycled back into the ecosystem. The vast trade in dried chips and pellets (more than 4 million tons from Thailand to the European Economic Community in 1978 alone) for livestock feed prevents nutrient recycling and has the disadvantages discussed in Chapter 12, although it is environmentally preferable to the use of crop land for livestock grazing.

Sweet and Irish Potatoes: Since sweet and Irish potatoes tolerate lower temperatures more successfully than do other tropical root crops, they are often grown on steep slopes at higher elevations under slash-and-burn cultivation system. While such crops have enabled peoples traditionally experienced with their cultivation to settle at higher elevations (Hongsapan, 1962), cultivation under mountainous conditions without special precautions can contribute to erosion and deforestation (Chapter 21). Creating contour ridges can prevent both erosion and the greening of the tubers.

Weed Control: As profitable cultivation of cassava requires at least two hand weedings, it is a very labor intensive and expensive operation. For high yields, most root crops need much weeding. Several pre- and post-emergence chemical herbicides can be used (de la Pena, 1979), but are less preferable environmentally than manual weeding (Chapter 22). Where the use of these chemicals is unavoidable, perhaps due to labor limitations, all appropriate precautions (Chapter 22) must be taken.

Pest Control: The presence of hydrocyanic acid (HCN or prussic acid) in a layer of tissue just under the skin of the cassava tuber reduces losses from pests. Cultivars with high HCN content (bitter varieties) often show greater net productivity than those with less HCN (sweet varieties). Bitter cultivars may also deteriorate less rapidly in storage than the sweet types.

Cassava hornworm has been biologically controlled by placing _Polistes_ wasps near the crop. The pest causing most concern in East Africa, the introduced green spider mite from South America,

- 31 -

damages up to 50 percent of the cassava crop. Two of its original predators--a beetle and another mite--were introduced and have already started to control the pest. Similarly, in Zaire, mealy bug control by original predators is encouraging (Chapter 18).

Food Processing and Storage: Since cassava is perishable, has a high water content (70-80 percent), bulky, heavy, and expensive to transport, it should either be stored where it is in the ground, or processed or consumed within a few days after harvest. Starch content, highest at about 18 months, deteriorates and lignifies only slowly in the ground. Local employment can be increased and village women released from onerous milling by small-scale processing of cassava into fermented gruel. Pilot facilities with an output of 1 ton per day have been successfully operated (Akinnele, et al., 1965); the nutrient value of the cassava is thereby improved.

Industrial Processing: Wastewater from industrial cassava starch extraction can create serious disposal and pollution problems. Effluent containing fine starch granules, colloidal material, and dissolved nutrients is high in biochemical oxygen demand (BOD), but may be used as a low-concentration fertilizer for irrigation. Once the hydrocyanic acid has been detoxified by fermentation, the effluent is useful as a growth medium for microbial protein production, or in fishponds. Large-scale processing sited downwind of habitation avoids the problem of foul odors. Biological oxidation plants for the effluent can produce sludge containing 35 percent protein. Alternatively, during starch extraction, the pulp can be fermented and, after detoxification, fed wet with the peelings to local livestock. Furthermore, it can be de-watered for transportation and used as livestock feed elsewhere. The pulp represents about 10 percent of the original tuber tissue and contains about 56 percent starch, 36 percent fiber, 55 percent protein, and 2-7 percent ash. Although all these processes reduce pollution problems, the economics of some require further testing to determine their commercial feasibility.

Ethanol Production: Ethanol production from cassava represents an environmentally benign, renewable conversion of free solar power into a high-grade fuel. To the extent that it does not decrease arable land or food calories available for human use, production of ethanol from cassava seems an appropriate source of fuel in tropical countries. If the mineral nutrients are recycled into the ecosystem, the system becomes more nearly sustainable. The alcohol yield per ton is greater than that for sugar cane (180 l/t vs. 67 l/t), but since cassava produces fewer tons per hectare per year (10-20 t/ha/yr vs 40-60 t/ha/yr), sugarcane produces more alcohol per hectare per year (3,015 l/ha/yr vs. 2,160 l/ha/yr). Sugar competes strongly with food crops on rich, moist soil, however, while cassava can be grown under widely varying and more marginal conditions (Brown, 1980).

Stillage produced during cassava-ethanol manufacture is voluminous (12 times the volume of alcohol) and highly polluting, but it can be recycled through pigs or poultry or used directly as a fertilizer, as is sugar stillage (Chapter 7.) The stillage can be used as a valuable substrate for the production of fungal protein, enzymes, antibiotics, and vitamins. When adequately air-dried, stems burned to provide process heat improve the distillation efficiency.

References

Akinnele, I. A., Cook, A.S. and Holgate, R.A. 1965. The manufacture of gari from cassava in Nigeria. New York, Gordon & Breach IV:633-644.

Araullo, E.U., Nestel, B. and Campbell, M. (eds.) 1974. Cassava processing and storage. Ottawa, IDRC: 125 p.

Bellotti, A., and Arias, B. 1977. Biology, ecology and biological control of the cassava hornworm. (227-232) in above.

Bennett, F.D. and Greathead, D.J. 1977. Biological control of the mealybug Phenacoccus manihoti Matile-Ferrero: prospects and necessity. Cali, CIAT, Cassava Protection Workshop: 181-197.

Brown, L. 1980. Food or Fuel: new competition for the world's cropland. Washington, D.C., Worldwatch Institute 35:43 p.

Cock, J. H. 1983. Cassava. Boulder, Colorado, Westview, 175 p.

CIAT, 1975, 1976, 1977. Abstracts on cassava. Cali, Colombia, CIAT, 3 volumes.

CIAT, 1976. Symposium of the International Society for Tropical Root Crops IV. Cali, Colombia, CIAT, 227 p.

Coursey, D.G. 1967. Yams. London, Longmans Green, 230 p.

de la Pena, R.S. 1979. Weed control in root crops in the tropics. (169-188) in Proceedings of the Symposium on Weed Control in Tropical Crops. Manila, Philippines, Weed Sci. Soc. Philippines, Inc.

FAO, 1977. Roots and Tubers. (Institut pour le Developpement Economique et Social, Abidjan, Ivory Coast.) Rome, FAO Better Farming Series 16:57 p.

Goering, T.J. 1979. Tropical root crops and rural development. Washington, D.C., World Bank, Staff Working Paper 324: 85 p.

Gomes, G. 1977. Life cycle swine feeding systems with cassava (65-71) in Cassava as Animal Feed. Ottawa, IDRC-095E.

Grace, M. R. 1977. Cassava processing. Rome, FAO Plant Production and Protection Series No. 3:155 p.

Hongsapan, S. 1962. Does planting of cassava really impoverish the soil? Kasiorn 35(5):403-407.

Ibekwe, G.O. 1978. Cassava bacterial blight: Abstracts of literature. Ibadan, IITA: 102 p.

IDRC, 1981. Tropical root crops: research strategies for the 1980's. Ottawa, IDRC, 279 p.

International Trade Center, 1977. Cassava export potential and market requirement. Geneva, UNCTAD/GATT: 65 p.

Jones, W.O. 1959. Manioc in Africa. Stanford, California, Stanford University Press, 315 p.

Kay, D.E. 1973. Root crops. London, Tropical Products Institute, Foreign and Commonwealth Office, ODA: 245 p.

KKU-IDRC, 1979. Annual report for 1978 and final report: Cassava Nutrition Project. Thailand, Khon Kaen Univ., Faculty of Agriculture.

Krochmal, A. 1966. Labour input and mechanization of cassava. World Crops 18(3): 78-80.

Leakey, C.L.A. and Wills, J.B. (eds.) 1977. Food crops of the lowland tropics. Oxford, Oxford University Press, 345 p.

Lozano, J.C. et al. 1976. Field problems in cassava. Cali, Colombia. CIAT Series GE-16:127 p.

Montilla, J.J. 1977. Utilization of the whole cassava plant: feed composition, animal nutrient requirements and computerization of diets. 1st International Symposium. Utah Agric. Experiment Station. Utah State Univ. Logan, Utah:98-104 p.

Nestel, B. 1973. Chronic cassava toxicity. Ottawa, IDRC: 162 p.

Oke, O.L. 1978. Problems in the use of cassava as animal feed. Feed Science and Technology 3(4):345-380.

Onwueme, I.C. 1978. The tropical tuber crops: yams, cassava, sweet potato, cocoyams. New York, Wiley, 234 p.

Persley, G., Terry, E.R. and MacIntyre, R. (eds.) 1977. Cassava bacterial blight. Ottawa, IDRC: 36 p.

Plucknett, D.L. (ed.) 1979. Small-scale processing and storage of tropical root crops. Boulder, Colorado, Westview, 461 p.

Scaife, A. 1968. The effect of a cassava "fallow" and various manurial treatments on cotton at Ukiriguru, Tanzania. East African Agricultural and Forestry Journal 33(3):231-235.

University of Georgia, 1972. A literature review and recommendations on cassava. Athens, Georgia, 326 p.

5
Coffee and Tea

When well-managed, both coffee (<u>Coffea</u> species) and tea (<u>Camellia</u> <u>sinensis</u>) are capable of protecting the soil against erosion, loss of structure, and desiccation. However, as these perennials are typically cultivated on sloping lands in high-rainfall areas, neglect of soil protection measures can result in drastic soil deterioration and loss. Inappropriate and excessive tilling was common during the early years of development of these crops, and this resulted in much damage to soils (Eden, 1976).

While coffee and tea contribute significantly to the quality of life in many communities, both rich and poor, neither crop offers nourishment to humans. Although production for local consumption is generally beneficial, large-scale cultivation of coffee and tea as a means of generating foreign exchange carries the inherent dangers--both financial and environmental--of over-reliance on any single luxury export. Coffee is the largest internationally-traded commodity (averaging US$12.2 billion during 1978-80).

Developing countries are now focussing on export diversification and self-reliance in food production as essential steps to reduce economic vulnerability and to promote economic self-determination and social well-being. Accordingly, other environmentally suitable crops, such as fruit and nut trees (or climbing perennial vines), can be seriously considered as alternative or supplementary crops. Projects designed to increase yields in areas already producing tea or coffee that include pilot plots of alternative taller perennial food crops will assist future diversification. Similarly, food-bearing trees can be chosen where shade is needed for the tea or coffee. The choice of species and management practices for such mixed stands create added opportunities for implementing an integrated pest management approach (Chapter 18).

<u>Coffee</u>: Coffee trees, whether <u>Coffea</u> <u>arabica</u>, <u>C.</u> <u>robusta</u>, <u>C.</u> <u>canephora</u>, or "Liberian" <u>C.</u> <u>liberica</u>), thrive on deep but well-drained loamy soils. To guard against erosion after

clearing, land sloping at more than 15 percent will probably need costly bench-terracing. Other sloping land can be protected by contour ridges and standard erosion control methods, as detailed in Chapter 25.

Mulching with material obtained locally is a useful general practice. This mulch is derived from stands of tall grasses (e.g., Napier grass) in valley bottoms, from intercropped bananas providing temporary shade, from nearby banana groves, or from composted coffee pulp and husks. Soil protection by mulching is preferable to intercropping with annuals, which inevitably compete with the coffee for moisture during the dry season. If mulch is applied at the recommended rate, supplementary fertilizer is often unnecessary. Mulching requirements for unshaded coffee will certainly be greater than for coffee that is grown in shade.

Shade trees are often grown over coffee (and tea) unless mists and cloud cover are regularly present. Where rainfall is low, shade trees are also dispensed with to avoid competition for water. The species commonly used provide fixed soil nitrogen (legumes), valuable timber (e.g., Grevillea robusta), firewood, or leaf mulch (Leucaena glauca), as well as the intended shade and protection against erosion. Experience over the years with trees to shade coffee has led to an excessively restricted list of species. However, additional research into the advantages of taxonomic diversity (improved nutrient status and pest reduction) and economic usefulness of shade trees would improve not only the yield and quality of the coffee, but would also benefit the soil and the economy.

Tea: Tea is grown on a wide range of soil types, usually well-drained, leached, and rather infertile. Much tea land has been converted from forest, the clearance of which is possibly the major environmental cost (Chapter 17). Careless clearing and subsequent inattention to erosion control measures frequently lead to rapid and irreversible loss of more fertile topsoil. To control erosion, wide-based ridges should be established, with tea shrubs planted along the contour as noted in Chapter 25. Tea prunings, along with fallen tea and shade tree leaves, help protect the soil and return nutrients when left on the ground.

The rate of removal of nutrients concentrated in plucked shoots is very high (e.g., 4-5 percent nitrogen (N), 1.5-2.5 percent potassium (K), and 1 percent phosphorus (P). Consequently, these nutrients must be replaced, either in organic or inorganic form. The rate of nitrogen fixation by leguminuous shade trees in Assam has been estimated at about 90 kg/ha/yr. Shade trees that are more deeply-rooted than the tea may absorb nutrients that percolate past the more superficial roots of tea and recycle them as their leaves drop, thus reducing fertilizer requirements.

Weed Management: Since coffee and tea both suffer from competition for N and water, it is worth expending great efforts before the crop is planted to eradicate weeds, particularly such tenacious, mat-forming types as stoloniferous or rhizomatous grasses. When the crop is young, intercropping with, for example, vigorously-growing cowpea may smother weeds; intercropping of annuals between rows also may control weeds but at the expense of extra labor and risk of erosion.

On established coffee and tea estates, the ideal condition would be to maintain a minimum growth of broad-leaved weeds or selected ground cover controlled by overhead shade, occasional weeding, and an annual mulching. Where mulching has been neglected, however, excessive clean-weeding has accelerated erosion in many areas. The magnitude of the weed problem is greatly influenced by differences in cultivation. Weed problems are increased by high fertilizer inputs and lack of overhead shade; they are diminished by growing the crop at a higher density under shadetrees. The environmental damage, including persistence and toxicity, of different herbicides is outlined in Chapter 17. Where biocide use is unavoidable, following precautions in their application will reduce environmental and health risks (Chapter 22).

Pest Management--Tea: There are severe limitations to the use of poisonous chemicals on a crop such as tea, whose leaves are harvested and consumed by humans. Even though mainly the younger shoots are plucked, the taint from chemicals can reduce the economic value of the tea crop and can sometimes impair health. For example, an aerial application of DDT to combat a locust invasion so contaminated tea harvests in Africa that entire crops were lost (Eden, 1976). A further danger in the use of biocides is that beneficial insects are also destroyed. For example, a recent spraying program in Sri Lanka against Helopeltis species was followed by an increase in the incidence of the tea tortrix pest, which probably resulted from a lethal effect of the biocide on the (introduced) Macrocentrus parasite of the tortrix larvae. As the full effects of biocides are usually more widespread than initially intended, there is much to be said for combatting insect pests of tea by improved methods of cultivation or biological means, rather than by resorting to large-scale biocide application.

Integrated pest management methods already have been successfully implemented for some important tea pests. Since tea is a perennial, its pests are suitable subjects for attempts at biological control by introduced agents. Thus the Macrocentrus parasite of the tortrix, introduced to Sri Lanka from Java, now provides an effective means of tortrix control there. The destructive nematode Pratylenchus pratensis can be controlled to a useful extent by intercropping with suitable marigold (Tagetes) species and by avoiding susceptible shade trees.

Several serious tea diseases (e.g., blister blight, root-splitting disease, and red root) have responded to non-chemical control methods. These methods rely on changing the timing of pruning, ring-barking and stump removal of unwanted trees, use of green manures, and manipulating shade intensity.

Pest Management--Coffee: Coffee pest management usually has been by chemical means, often relying on the obsolete, broad-spectrum, and persistent biocides aldrin, dieldrin, and endrin. The Environmental Protection Agency (EPA) now has banned these biocides for most uses within the United States. To lessen dependence on polluting substances, integrated control methods against coffee pests should be developed and introduced as soon as possible. Indeed, biological control should be a major focus of research associated with agricultural projects involving perennial crop systems. Diseases in coffee, especially leaf rust, will remain serious threats to production until resistant coffee varieties are developed. In the meantime, use of biocides against these diseases is likely to continue indefinitely. Integrated pest management techniques cannot be simply transferred; they must be developed separately for each agro-environment. For this reason, local research linked to local agricultural projects is particularly necessary (Chapters 17 and 18).

Coffee and Tea Processing: Under efficient operating conditions, there is relatively little waste from the processing of tea and coffee, and hazardous pollutants are not normally present. However, in some countries, safety standards in processing plants are "virtually non-existent" (WDM, 1979) and water pollution is serious in places. Coffee processing centers accumulate piles of solid wastes (skins, hulls, and stalks) that encourage flies and become smelly. These solids can be used as mulches. The Institute of Nutrition of Central America and Panama and the Canadian International Development Research Center have shown that coffee pulp has the potential to yield nutritious food for farm animals and fish (Braham and Bressani, 1979), as well as single-celled protein, ethanol, vinegar and biogas (Adams and Dougan, 1981). Coffee production will be more socially beneficial in countries that undertake such processing (WDM, 1979).

Fuel for Drying and Curing: Deforestation, intensified by the need for fuelwood to dry and cure coffee, tea and tobacco, is an environmental cost outlined in Chapter 12. To the extent coffee and tea are grown in wetter areas than tobacco, and are more solar-dried, their pressure on the fuelwood supply (Chapter 8) is likely to be less.

References

Adams, M.R. and Dougan, J. 1981. Biological management of coffee processing wastes. Tropical Science 23(3):177-196.

Braham, J. E., and Bressani, R. (eds.) 1979. Coffee pulp: composition, technology and utilization. Ottawa, IDRC, 108e. 8(1):95 p.

Branon, R.H. 1964. Coffee: a background study with primary emphasis on Guatemala. Madison, Univ. of Wisconsin. Land Tenure Center, 51 p.

Buzzanelli, P.J. 1979. Coffee production and trade in Latin America. Washington, D.C., USDA:82 p.

Eden, T. 1976. Tea. London, Longmans, 236 p.

Haarer, A.E. 1963. Coffee growing. London, Oxford University Press. 127 p.

Haarer, A.E. 1962. Modern coffee production. London, Leonard Hill, 495 p.

Harder, C.R. 1964. The culture and marketing of tea. Oxford, Oxford University Press, 262 p.

Krug, C. 1968. World coffee survey. Rome, FAO: 476 p.

Le Pelley, R.H. 1968. Pests of coffee. London, Longmans Green, 590 p.

McFarquhar, A.M.M. and Evans, G.B.A. 1972. Employment creation in primary production in less developed countries: case studies of employment potential in the coffee sectors of Brazil and Kenya. Paris, Organization for Economic Co-operation and Development:1652 p.

Sarkar, G.K. 1972. The world tea economy. Delhi, Oxford University Press, 237 p.

Singh, S. 1977. Coffee, tea and cocoa: market prospects and development lending. Washington, D.C., World Bank, Staff Occasional Paper 22:129 .

UNDP, 1978. Coffee: Co-operation on a vulnerable commodity. New York, UNDP, TCDC Case Study No. 11:14 p.

Wellman, F.L. 1961. Coffee: botany, cultivation, and utilization. London, L. Hill, 488 p.

Wickizer, V.D. 1951. Coffee, tea and cocoa: an economic and
    political analysis. Stanford, California, Stanford University
    Press, 497 p.

World Bank, 1979. Environmental guideline on tea and coffee
    production. Washington, D.C., Office of Environmental Affairs,
    World Bank. 5 p.

WDM, 1979. The tea trade. London, World Development Movement, 44 p.

6
Cocoa

The cocoa (<u>Theobroma</u> <u>cacao</u>) tree's natural environment is the lower story of evergreen neotropical rain forest; much cocoa is produced in hot, humid tropical rain forest areas that have little or no distinguishable dry season. Cocoa, a small tree with a relatively closed canopy when grown in monoculture, exhausts tropical soils less than annual cropping does. Furthermore, cocoa is grown often with shade trees, frequently leguminous species (e.g., <u>Gliricidia</u>, <u>Leucaena</u>), food trees such as coconuts, or (during the first two to three years after planting) large herbaceous perennials such as bananas, which also protect soils from erosion. However, large-scale clearing of tropical forests for cocoa production reduces the habitat of thousands of plant and animal species and can irreversibly destroy these unique ecosystems (Chapter 12). Such major losses can be avoided by improving yields on land already under cocoa cultivation, rather than by expansion of cocoa cultivation at the cost of tropical forest. Where a country is involved in settling populations in tropical forest areas, such as in the Amazon Basin, cocoa cultivation is environmentally preferable to annual cropping or livestock production, although local food needs must have priority.

While cocoa and associated shade trees protect soils more than annual crops do, continued cultivation can result in gradual deterioration of soil through the loss of organic matter and nutrients. Mulching and growing other tree species within cocoa plantations, as is practiced in West African "compound" farms, helps compensate for nutrient loss. Undisturbed soils under natural forest in the humid tropics usually possess a relatively thick surface layer of crumb-structured material which is rich in organic matter and plant nutrients. This layer undoubtedly contributes to the success of cocoa planted underneath natural forest, or on land which has been recently cleared from forest. However, forest clearing and cultivation lead to the rapid destruction of the valuable crumb layer. Protection of the crumb layer is an important aspect of soil management in cocoa husbandry because the plant itself is shallow-rooted. Mulching with cocoa pod husks, pen manure, sawdust, bagasse, and grassy materials is an effective way to enhance the soils.

Pest Management: The common usage of biocides, along with their misuse, creates problems. For example, mirids, major cocoa pests, were traditionally controlled to some degree by leaving young cocoa in "bush", but there was no solution to attacks on adult cocoa. Control seemed possible with the advent of biocides; DDT was used extensively in the 1940's, as was lindane during the 1950's. However, mirid resistance to the biocides became widespread within a decade in major cocoa-producing areas, especially in Africa. For example, mirids in a cocoa-growing region were found to be more biocide-resistant than mirids from a non-cocoa-growing region by the following factors: aldrin (over 1,000 times); endrin (over 10,000 times); dieldrin (over 1,000 times); thiodan (100,000 times); heptachlor (15 times); and chlordane (100 times).

Pest outbreaks frequently worsened in Malaysia after the onset of biocide sprayings because of the damage to natural pest enemies:

> "It is significant that ... in Sabah, where (contact-acting insecticides) were used much less, the major pest outbreaks did not occur, although there has been damage from the branch borers and from (some) low-density pests ... (Elsewhere) cessation of the spraying at the end of 1961 appears to have corrected the situation. The major outbreaks have come under natural control, leaving only a few pests, damaging at low densities, that required selective, artificial control measures. As the cocoa has grown older and the micro-habitats have changed, other pests -- as was to be expected -- invaded the crop. But the absence of contact acting insecticides has meant that, for the most part, these have come quickly under effective natural control, leaving again only the low-intensity pests in need of attention." (Conway, 1972).

The biological control of cocoa pests should be an important component in all projects, if chemical control - with all its hazards, costs, and severe and ultimate limitations - is to be reduced to a minimum. Research components of development projects should therefore accord biological control a high priority. Where use of chemical controls is currently unavoidable, all precautions must be taken not only to avert threats to human health, but also to limit environmental contamination (Chapters 17 and 18).

Drying and Storage: The sun-drying of cocoa in easily constructed barns with moveable roofs is feasible in many regions and is to be preferred as a means of saving on fuelwood or other fuel. Pest and disease attack of stored cocoa can be prevented with well-designed storage barns, good ventilation and strict hygiene, rather than by using biocides.

Utilization of By-products: Cocoa by-products can be utilized productively, rather than disposed of in ways that cause environmental problems. For example, Ghana's present annual dry cocoa bean production of about 400,000 tons is accompanied by the production of 4 million tons of fresh cocoa-pod husk material that either can be fed directly to animals or formulated into animal feed and profitably recycled. Furthermore, pod husk material contains extractable pectin of reasonable quality.

Social and Economic Considerations: A number of African nations with comparative advantages for cocoa production rely on the crop for much of their foreign exchange. However, over-reliance on one commodity for national income has led to economic vulnerability in years of poor harvest, and uncertainty during world price fluctuations. It is common for extensive new plantings to be established during times of high world prices, only to be affected by a decline in prices when the new areas come into production several years later. Developing countries seeking a measure of economic self-determination and well-being frequently deem it essential to diversify exports and to promote a degree of self-reliance in food production. In view of current (1983) 40 year lows in most commodity prices, countries without these policies can become vulnerable to international market forces over which they can exert no control. The conversion of land from domestic food production to export crop production can expose already weak developing country economies to unacceptable levels of risk. The fact that cocoa is a luxury product may make it especially sensitive to demand fluctuations, as demand for such items as chocolate in importing countries can drop considerably during times of economic recession.

References

Are, L. and Gwynne-Jones, D.R.G. 1974. Cacao in West Africa. Ibadan, Oxford University Press, 146 p.

Conway, G.C. 1971. Pest of cocoa in Sabah and their control: with a list of cocoa fauna. Kota Kinabalu, Sabah, Malaysia, Ministry of Agriculture and Fisheries, 125 p.

Conway, G. C. 1972. Ecological aspects of pest control in Malaysia. (467-488) in Farvar, M.T. and Milton, J.P. (eds.) The Careless Technology. New York, Natural History Press, 1030 p.

Entwistle, P.F. 1972. Pests of Cocoa. London, Longmans Green, 779 p.

Evans, H.C., Edwards, D.F. and Rodriguez, M. 1977. Research on cocoa diseases in Ecuador: past and present. PANS 23(1):68-80.

Mainstone, B.J. 1978. Crop protection in cocoa in Malaysia. Planter 54(631):643-658.

Ramadasan, K., Addullah, I., Teoh, K.E., Vanialingam, T., and Chan, E. 1978. Intercropping of coconuts with cocoa in Malaysia. Planter 54(627): 329-342.

Simmons, J.L. (ed.) 1976. Cocoa production: economic and botanical perspectives. New York, Praeger, 413 p.

Singh, S., de Vries, J., Hulley, J.C.L. and Yeung, P. 1977. Coffee, tea, and cocoa: market prospects and development lending. Baltimore, Johns Hopkins University Press, 129 p.

Thorold, C.A. 1975. Diseases of cocoa. Oxford, Clarendon Press, 423 p.

Wickizer, V.D. 1951. Coffee, tea, and cocoa: an economic and political analysis. Stanford, Stanford University Press, 497 p.

Wood, G.A.R. 1975. Cocoa. London, Longmans Green, 292 p.

7
Sugar Cane

Sugar cane (<u>Saccharum</u> <u>officinarum</u>), is a highly efficient converter of solar energy. In certain sites, it is the world's most productive crop, producing as much as 8,000 kcal/m^2/year in Hawaii. Under good management on fertile soils, cane yields can exceed 100 t/ha/year, and have risen to as much as 220 t/ha/year under experimental conditions. Since sugarcane is a perennial and is usually grown on gentle slopes or flat lands, erosion is usually not a major problem. The plant's extensive rooting capabilities and growth even under relatively adverse conditions make it an attractive crop; environmental concerns derive more from its processing than from its cultivation.

<u>Impact of Sugar Cane Cultivation on Soil and Water</u>: While mostly a rainfed crop in the humid tropics, sugar cane is particularly productive when irrigated under the high light intensities of arid and semi-arid tropical and sub-tropical areas (e.g., Egypt, India). A trash layer formed of sugar cane tops and leaves helps to conserve soil moisture, control weeds, and improve soil organic content, particularly in seasons of low rainfall. Soil organic matter levels are raised even more if root renewal is stimulated by ratooning.

Since the possibility of water-related diseases is increased by the implementation of irrigation projects, such projects require health components to control diseases, particularly schistosomiasis and malaria (Chapter 20). Run-off from cane fields may contain undesirable concentrations of fertilizers and biocides. It is therefore important to consider downstream water uses before project implementation to ensure compatibility (Chapters 15 and 20).

Since industrial yields of 15-100 tons/ha of sugar cane remove considerable amounts of nitrogen (N), phosphates (P), magnesium (Mg), calcium (Ca), and potassium (K) from soils, fertilizer application is required if such yields are to be sustained. However, because the product (sucrose) is oligotrophic (low in minerals), artificial fertilizer needs can be reduced to the extent that filter press-mud (rich in P, K, Ca, Mg), mill water, ash, and factory residues are recycled into the cane fields. Since nitrogen, sulfur and possibly even phosphorus are volatilized during preharvest burns, the benefits of burning may decrease as fertilizer costs rise. Chopping, ensilage and composting are becoming feasible, as in Zaire's Kwilu Sugar Project.

Bagasse: By far the most important use of bagasse is to generate steam. Because disposal of excess bagasse can be expensive or polluting, factory operations should be adjusted, where feasible, to balance the use of bagasse with its production. The best use of bagasse - boiler fuel - will be favored to the extent fossil fuel prices continue to rise. Reliable amounts of excess bagasse production are increasingly being recycled in various ways. For example, bagasse can successfully substitute for several wood products, including fiberboard and paper. Bagasse also can be used to produce single-cell protein, biogas and fertilizer sludge. Using mechanical separators to sort the wax, rind and pith before crushing facilitates the productive uses of these by-products. With supplements, decorticated pith can be used as cattle feed. In some projects, excess bagasse is use as compost.

Weed Management: Weeds compete seriously with young cane for water and nutrients until the growing cane shades them out. With fertilizer supplements, vegetables or other annuals can be grown to suppress weeds; the economic benefits of such short-term intercropping usually outweigh the minor losses to the sugar crop from competition with the annuals. Where herbicide use is unavoidable because of labor limitations, the precautions noted in Chapter 22 will minimize harm.

Pest Management: A major drawback of any monoculture of succulent plants, including sugar cane, is the risk of pests and diseases that can attack the crop at all stages of growth. The advantages of the trash layer in lowering the consumption of nutrients and water, may be partly offset to the extent that pests and diseases are harbored. The feasibility and safety of all chemical controls must be verified and precautions observed during usage. For example, chlordane, one of the biocides previously much used to treat cane cuttings but now declared unsafe, is being phased out completely in the United States for all uses except surface termite control. The height and density of sugar cane makes effective ground-level application of biocides both difficult and hazardous (Chapter 17).

Due to the human health, environmental, and biological disadvantages of biocide use, integrated pest management and non-chemical insect control techniques are strongly preferred whenever practicable. Much information has become available on disease resistant crop varieties, cultural methods of pest control, biological, mechanical, and chemical control methods, but the methods have yet to be integrated in many cases. However, the sugar cane agro-ecosystem, more permanent than those of annual crops, theoretically provides better possibilities for use of natural predators in a system of integrated pest management. For example, control of stemborers by release of artificially-reared predatory insects is now commercial practice in many regions.

As outlined in Chapters 17 and 18, biological control is a major thrust of research components associated with agricultural projects. Integrated pest-management techniques cannot be simply transferred; they must be developed individually for each type of agro-ecosystem. Local integrated pest management research linked to sugar cane projects is particularly needed.

Harvesting: The burning of leaves of standing cane greatly facilitates manual cane cutting and harvesting. It also reduces moisture content and increases cane sucrose content, provided that harvest immediately follows burning. If repeated for years, however, the resulting loss of organic matter may adversely affect the soil and its nitrogen and sulfur levels. Plantations sited downwind of settlements can avoid the health hazard and environmental nuisance of smoke from burning cane. Failing such siting, burning should be carried out during appropriate wind conditions to the extent feasible. The extremely seasonal labor requirement occasioned by manual harvesting often implies seasonal under-employment. Moreover, sudden population influx during harvest requires special public health measures in order to prevent outbreaks of malaria and other diseases brought in by migrant laborers. During this period, adequate and safe water supplies and provisions for waste disposal are also essential.

Sugar Production and Refining: Potential pollutants from sugar factories include smoke, cane washing water, factory waste water in general, filter-press muds, and bagasse and molasses (where not recycled). Cane washings and factory wastes contain inorganic solids and some sugar and other organic matter; their Biochemical Oxygen Demand (BOD) is usually not high. Filter muds, however, contain large amounts of organic material (particularly sugars), and are commonly recycled, given away, or even sold. These wastes are high in BOD and have caused severe pollution where improperly handled. Discharging high BOD wastes into settling ponds, rather than dispersing then into water-courses (where it is lost to the source ecosystem), reduces pollution. Pond sediments, relatively such rich in calcium and other nutrients, can be periodically recycled on the cane fields. Bagasse burned as fuel in the refinery's boilers provides minerals in the ash that reduce fertilizer requirements when recycled on land (as in nurseries), assuming toxic silica buildup can be avoided. If used instead in road construction or other non-agricultural uses, the nutrient value of the ash may be underutilized. Smoke from the sugar refinery can cause air pollution. Smokestacks situated downwind of populated areas and high enough to allow adequate smoke dispersal can protect workers and local inhabitants from smelly fumes.

Ethanol Production: Fermentation and distillation of cane sugar and molasses into ethanol is environmentally desirable, in that it represents sustained-yield conversion of solar energy.

Ethanol, high-grade fuel or raw material for the chemical industry, is thereby produced from a fully renewable resource. Cane sugar can be the basis for as wide a variety of chemicals as are now derived from petroleum. Clearly, where sugar is required for food or the land is needed for other food crops, human nutritional tradeoffs have to be addressed (Chapters 4 and 20). Under current practices, the social impact is severe. Large cane growers in Brazil, for example, purchase land from small food growers. This increases the price of agricultural land and forces small farmers to more marginal and distant land. The result can be higher food prices and inflation. Furthermore, the amount of useful ethanol energy produced only slightly exceeds the amount of energy consumed in production. In a large (50 million gallons per year) modern factory, about 70,000 BTU are required to produce one gallon of ethanol containing only 76,000 BTU. If a new technology can remove the 8 percent ethanol from the 92 percent water, only then will real progress be made (cf Pimentel, et al., 1983).

The three main byproducts of cane fermentation and distillation are carbon dioxide, fusel oils (or feints), and stillage. Carbon dioxide, fixed from the atmosphere by the cane during photosynthesis, is returned to the atmosphere during fermentation. Fusel oils are usually collected from the rectification column and sold (World Bank, 1980). Since stillage, vinasse or lees (the residual liquor remaining after fermentation and distillation), is voluminous (13 times the volume of alchohol produced), initially very hot, corrosive (approximately pH4), highly polluting (BOD=10,000 to 30,000 mg/lt; cf domestic waste = 300 mg/lt), its improper disposal (in the ocean, or in rivers) massively impairs water quality. A medium size plant produces 120,000 liters of stillage per day. A large plant produces one million liters daily, which is potentially more pollutive than the sewage of an entire city. Since stillage contains about 3 percent nutrients (including 1.5 percent potassium, 1 percent nitrogen, 0.2 percent phosphate), its direct recycling onto fields boosts yields, if the fields are close enough (e.g., 3 km) to the distillery for transport costs to be economical. Combining stillage with flood fallow is occasionally useful and would improve fish yields. Stillage sprayed on cane compost improves the product, but the compost can absorb only limited volumes of stillage. Distiller's grains, produced by evaporating stillage, are sold for animal feed. However, since stillage is so dilute (10 percent solids), evaporation costs are high.

Anaerobic digestion of stillage produces ten liters of gas (60 percent methane) per liter of stillage. The main constraint is the digestion time (20 days). A 10,000 liter capacity digester can handle only 500 liters of stillage per day, while still leaving some residue. Research in progress to speed and improve this process is encouraging. Algae can be grown in some stillage ponds and fed to pigs. Chemical and physical treatment (flocculation, filtration, sedimentation, reverse osmosis) is used in special cases, but is too expensive for general use at present.

References

Anwar, A.A. 1971. Production of sugar: policies and problems. Lahore, Board of Economic Inquiry, Punjab Publication No. 148:397 p.

Ballinger, R.A. 1971. A history of sugar marketing. Washington, D.C. USDA, Foreign Agricultural Service, 126 p.

Barnes, A.C. 1974. The sugar cane. New York, Wiley, 572 p.

David Livingstone Institute of Overseas Development Studies. 1975. A report on a pilot investigation of the choice of technology in developing countries. Glasgow, 121 p.

Development and Resource Corporation. 1978. An update on world sugar supply and demand, 1980 and 1985. Washington, D.C., USDA, Foreign Agricultural Service, 30 p.

Forsyth, D.J. 1979. The choice of manufacturing technology in sugar production in less developed countries. London, HMSO, 116 p.

Grissa, A. 1976. Structure of the international sugar market and its impact on developing countries. Paris, Organization for Economic Cooperation and Development, 120 p.

Johnson, D.G. 1974. The sugar program: large costs and small benefits. Washington, D.C., American Institute for Public Policy Research, 90 p.

Monypenny, R. 1976. The sugar beet and cane industries: a selective economic bibliography. Townsville, Australia, CSIRO, 110 p.

Paturau, J. 1969. By-products of the sugar cane industry: an introduction to their industrial utilization. New York, Elsevier, 274 p.

Perston, T.R., and Leng, R.A. 1978. Sugar cane as cattle feed. Part 1. nutritional constraints and perspectives. Centro Dominicano de Investigacion Pecuaria y Cana de Azucar. World Animal Rev. (FAO) 27:7-12.

Pimentel, D. et al. 1983. Environmental and social costs of biomass energy. BioScience (in press).

USDA, Foreign Agricultural Service. 1976. Sugar: world supply and distribution 1954/55-1973/74. Washington, D.C. USDA, 50 p.

World Bank, 1979 a. Cane sugar: mill and refinery operations: occupational safety and health guidelines. Washington, D.C., World Bank, Office of Environmental Affairs, 4 p.

World Bank, 1979 b. Cane Sugar: agricultural operations: occupational safety and health guidelines. Washington, D.C., World Bank, Office of Environmental Affairs, 3 p.

World Bank, 1980. Alchohol production from biomass in developing countries. Washington, D.C., World Bank: 69 p.

8
Tobacco

"The true 'wealth of nations' is the health of the individuals."

--Irving Fisher

Tobacco (<u>Nicotiana</u> <u>tabacum</u>) production creates a particularly difficult dilemma for development. On the one hand, tobacco projects may be highly profitable for small farmers, for modern plantations, and for governments. Tobacco may generate significant foreign exchange, as well as increasing rural employment. Tobacco production can also satisfy local demand and displace expensive imports. Tobacco is the most profitable crop for some regions, and from small family plots. On the other hand, the damage to public health and to the environment in the long term appears substantially to outweigh the benefits. This chapter outlines the dilemma, and presents options for resolving it.

Economic Importance: Leaf tobacco is now grown in virtually all countries except those in northern Europe. An estimated 5.66 million metric tons of leaves (farm sales weight) were produced in 1982, valued at US$12.7 billion. The foreign exchange earned from leaf exports (Figures 8.1 and 8.2) is economically significant to some developing countries. Just over half (52 percent) of the tobacco produced is consumed in the country of origin. Raw tobacco accounts for 1.5 percent by value of total world agricultural exports. Although 55 percent of leaf exports originate in developing countries, these contries have a very small share in tobacco manufactures. Leaf exports account for about 10 percent of Turkey's exports, for example, and over 55 percent of Malawi's--a large 8 percent of its Gross National Product.

Figure 8.1: <u>Tobacco Production in Some Developing Countries</u>

<u>Production Rank</u> (1980)	<u>Country</u> (over 10,000 MMT per year)	Leaf <u>Production</u> (1,000 MMT, 1980)	<u>Area Harvested</u> (1,000 ha, 1980)	<u>Unmanufactured Exports</u> (MMT)	<u>Value</u> (US$1,000 1980)
1.	P. Rep. China	920	708	33,000	74,100
2.	Brazil	410	309	143,555	290,036
3	India	400	450	73,179	150,867
4.	Turkey	230	260	83,727	233,742
5.	Zimbabwe	114	62	92,951	182,749
6.	Greece	113	91	69,633	213,560
7.	Korea Rep.	92	46	33,648	83,977
8.	Thailand	86	152	39,562	66,371
9.	Indonesia	82	165	28,339	58,848
10.	Pakistan	78	50	370	763
11.	Mexico	72	51	24,168	48,325
12.	Argentina	64	53	17,402	30,000
13.	Philippines	60	75	20,369	28,819
14.	Malawi	58	85	63,090	129,445
15.	Yugoslavia	56	58	24,953	84,063
16.	Burma	55	62	-	-
17.	Colombia	52	36	15,000	24,837
18.	Dominican Rep.	49	30	20,907	33,328
19.	Bangladesh	44	50	-	-
20.	Romania	40	50	8,000	20,000
21.	Spain	30	17	7,490	10,707
22.	Viet Nam	22	28	-	-
23.	Venezuela	21	16	320	323
24.	Tanzania	18	36	10,000	29,000
25.	Iran	15	13	-	-
26.	Paraguay	14	12	14,858	10,142
27.	Nigeria	13	31	-	-
28.	Syria	13	13	2,200	7,500
29.	Portugal	12	6	-	-
30.	Iraq	11	12	-	-
	World	5,386	4,284	1,369,136	3,859,465

Sources: FAO, 1981; USDA, 1982.
MMT = million metric tons

Figure 8.2: Value of Tobacco Exports as a Percent of Merchandise
             Exports

Rank	Country	Percent
1.	Malawi	55.6
2.	Zimbabwe	15.3
3.	Turkey	10.3
4.	Greece	5.54
5.	Tanzania	5.54
6.	Dominican Republic	4.05
7.	Paraguay	3.32
8.	India	2.16
9.	Brazil	1.90
10.	Thailand	1.26
11.	Yugoslavia	1.24

Sources: Unmanufactured export tobacco value from FAO (1981);
            merchandise exports from World Development Report
            (1981); only developing countries with over 1 percent
            of their merchandise exports as tobacco are included.

Since leaf production remains overwhelmingly labor-intensive, tobacco generates much employment. Tobacco taxes are also a source of revenue to most governments. Muller (1978) calculates that 4.7 percent of the Philippine Government's revenue accrues from tobacco taxes. Such taxes generally exceed the sale value received by the farmers. Taxes may account for between 30 and 50 percent of the retail price. Since the elasticity of demand for cigarettes is relatively low, historic price increases appear not to have been sufficient to decrease long-term consumption. Therefore, tobacco production greatly and directly benefits farmers (through sales and jobs), governments (through tax revenues, employment, and foreign exchange) and the manufacturers, whereas the costs are less direct, delayed, and spread throughout much of the population.

Inelastic Consumer Demand: To the major supply-side benefits mentioned are added the only benefit perceived by users: mild but transient pleasure. Even suppression of hunger now is being questioned. However, since tobacco is a habit-forming (or addictive) narcotic drug, most smokers are unable to smoke cigarettes in a controlled, non-dependent manner. Hence the high inelasticity of demand, particuarly in rich addicts. Tobacco is more likely to be equated in smokers' expenditure decisions with such necessities as food and shelter, rather than with luxury items (Shepherd, 1975). More than 1 percent of disposable personal income is spent on tobacco products in the United States (USDA, 1983).

Health Effects of Smoking: The health hazards of tobacco were contended almost from its first use (e.g., King James I "counterblaste" of 1604) and suspicion grew thereafter. Three factors combined to confirm this suspicion in retrospect, and to determine the magnitude of the problem. All forms of tobacco--even snuff--have now been found to be harmful, but tobacco consumed as cigarettes is more dangerous than other forms. Cigarette smoking is relatively recent. It became widespread only from the 1920-1940 period, at different rates in different countries and among men earlier than in women. Now a high and rising proportion (about 85 percent) of world tobacco production is consumed as cigarettes (EIU, 1980). Many serious health effects of smoking are not immediate. A latency period separates cause and effect: in the case of cancers, it can be two decades or more. This latency is shorter in the case of non-cancerous diseases, such as heart disease and stroke, and even shorter in pregnancy (spontaneous abortion), and in damage to infants.

Surprising as it may seem to those following the unfolding smoking-health evidence in recent years, apparently many people remain unconvinced of the damage caused by smoking. Gratification of immediate needs outweighs considerations of potential long-term health damage. Even consumption of foods like sugar and meat in excess will, by definition, damage health. Tobacco is distinctive in its effects because it is not a food; the threshold for health damage is low; the risk of damage is proportional to consumption and is cumulative; consumption is addictive; and health damage accrues to the consumer, family, neighbors, fetuses, and children.

Tobacco contributes more heavily to the number of cancer deaths than any other known substance. In the United States alone, tobacco causes more deaths than any other substance--and greatly more then those caused by auto accidents and excessive alcohol consumption. The "suggestion" of a cigarette-smoking/lung cancer link during the 1920s became a "confirmed association" in the 1950s, and a "causal relation" in the United States Surgeon General's Report of 1964 (OTA, 1981). Since that time, thousands of studies have confirmed and refined our understanding of smoking-related effects. Lung cancer is the most frequent cancer caused by tobacco smoking, but smoking is also associated with cancer at many other body sites, including larynx, oral cavity, esophagus, bladder, kidney, and pancreas. Increases in cardiovascular disease and strokes are even more widespread results of cigarette smoking than are cancers. Smoking increases both the risks of acquiring influenza and its severity (Kark, 1982). Chronic bronchitis, emphysema, and adverse effects on pregnancy also are serious smoking-related problems.

Cigarette smoking also damages pregnancy and increases infant mortality, a major public health concern worldwide and a major indicator of the overall health of a nation. Fourteen percent of all preterm deaths are attributed to maternal smoking in the US (Surgeon General, 1979). Women may not yet be smoking in large enough numbers in many developing countries for an increase in infant mortality from smoking to be apparent. However, smoking rates among females there are rising, as they have done in developed countries. Although cigarette consumption represents dependence (addiction), the tobacco industry invests in major campaigns to manipulate women towards association of cigarette smoking with independence, emancipation, and freedom. Women are fast gaining equality with men in cigarette-induced deaths, and girls are overtaking boys in taking up smoking. Even cognitive skills are impaired in children of smoking mothers, according to Socol, et al. (1982).

Lung cancer risks may be reduced somewhat by switching to low tar and possibly to low nicotine cigarettes. But for cardiovascular disease deaths, which is the largest component of excess mortality due to smoking, there is no evidence at present of such a reduction in risk. Evidence of less severe risk also is lacking for the other major health effects of smoking, notably chronic obstructive lung disease and effects on pregnancy. In fact, the number of cigarettes smoked may be more important then their tar, carbon monoxide, and nicotine content, partly due to the presence of radioactivity in cigarette smoke (NEJM, 1982; Kaufman, 1983). The radiation dose received by a one and a half pack per day smoker is the equivalent of 300 chest X-rays a year. A still unresolved question about low tar and nicotine cigarettes is the possibility of new risks being introduced through changes in design, filtering mechanisms, tobacco ingredients, or additives.

Many developing countries lack well-established vital statistics systems, making it difficult, if not impossible, to detect tobacco-related increases in mortality. However, various studies using subgroups of the population in Indonesia, Papua New Guinea, India, South Africa, Soviet Union, Zimbabwe, and Cuba (Sutnick and Gunawan, 1982; Dutta-Choudhuri, et al., 1959; Gelfand, et al., 1968; Cooper, 1982; Scrimgeour and Jolley, 1983) all reveal the link between smoking and disease. No population is known to be immune from the adverse health effects of tobacco. The severity of cigarette smoking in developing countries is emphasized by the United Nations World Health Organization (1975):

> "Smoking-related diseases are such important causes of disability and premature deaths in developing countries that (at least for children over five years) the control of cigarette smoking could do more to improve health and prolong life in these countries than any single action in the whole field of preventive medicine."

As aggressive promotion and consumption of cigarettes in the developing world continue to increase, so does the threat of an epidemic of smoking-related illnesses and deaths. Recent restraints on cigarette promotion in developed countries are said to make manufacturers seek expanded markets in developing countries where such restraints are lacking. In Africa, for instance, the per capita cigarette consumption has increased by 33 percent over the last 10 years (Anderson, 1981). Cigarette smoking doubled between 1967 and 1976 in Libya and Ethiopia. Furthermore, evidence is accumulating that tar levels of cigarettes sold in developing countries may be twice the levels of the same brands sold in developed countries. Without resolute action, WHO (1979) warns that "... smoking diseases will appear...before communicable diseases and malnutrition have been controlled."

Tobacco production and consumption have an economic "opportunity cost," harming human health in at least two other ways. In low-income families, although relatively less money is spent on cigarettes, even this decreases potential expenditures on food. Smokers worldwide spend between US$85 billion and US$100 billion annually to buy four trillion cigarettes (Eckholm, 1978). Moreover, if land or labor is scarce, their use for tobacco cultivation will reduce their availability for food production. To the extent that cash earned from such tobacco sales is less applied to buying food, the nutritional state of the poor will further decline. Reduced local food production may lead to higher prices, which penalizes even non-smoking families.

Tobacco, especially in processed form, that is profitably exported from a developing country generates valuable foreign exchange with little or no damage to the exporting country's health. However, since most tobacco is consumed in the country of origin, the benefits are reduced by the costs of damaged health. It may be argued that smokers are a self-selecting group exercising personal choice which does not affect others. This is not the case for at least two reasons. First, in most countries, the costs of impaired health of the smoker are borne more by the society (particularly the government and the taxpayers), and less by the smokers. Second, much smoking occurs in the presence of others who may not be able to restrict their involuntary or passive smoking (inhalation of smoke-polluted air), which can also damage their health (Shephard, 1982).

Soil Degradation: Tobacco requires either fertile soils or regular inputs of commercial fertilizer. However, most tropical soils are characterized by low nutrient content, particularly by deficiencies of phosphorus, often nitrogen, sometimes potassium. Tobacco production therefore usually depends on commercial fertilizers, the prices of which (especially nitrogen) are rising so sharply that they are increasingly out of reach of most developing country farmers. Furthermore, tobacco depletes soil nutrients at a much higher rate than many other crops, thus rapidly decreasing the life of the soil (Figure 8.3). An alternative to dependence on commercial fertilizer is to exhaust soil fertility in one or two years, then to clear new land (often by deforestation) for another plot.

Figure 8.3: <u>Depletion of Soil Nutrients by Tobacco and Other Crops</u>
(loss in kg/ha)

Harvest of one ton per ha	Nitrogen N	Phosphorus P_2O_5	Potassium K_2O
Tobacco	24.4	14.4	46.4
Coffee	15.0	2.5	19.5
Maize	9.8	1.9	6.7
Cassava	2.2	0.4	1.9

Source: Van Wambeke, 1975.

Biocides: Tobacco is one of the crops on which biocides are used most heavily. Vast quantities of biocides (see Chapter 17) are applied to tobacco crops virtually throughout its seven to eight month growing season. Most of these biocides are toxic; some are carcinogenic. Besides being hazardous to users, these chemicals can contaminate village water supplies. Although most western governments either ban or severely restrict the use of obsolete persistent organochlorine biocides, their use in developing countries continues. For example, aldrin is widely supplied by the international tobacco monopoly in Kenya (Madeley, 1982). Warnings are printed in two of Kenya's 15 languages (Swahili and English). Even if the user can read and understand the warnings, it is not easy to "avoid contaminating rivers", nor to "wash with soap and water after use". Most users have never seen a physician, and certainly are not able to consult one "immediately" as advised on the label.

The Fuelwood Crisis: Although much tobacco is sun-cured, the remainder involves using large amounts of wood. Where wood is used, curing is highly energy intensive, accounting for 10-15 percent of the product price in the case of Thailand (Schramm and Munasinghe, 1981). The Economist Intelligence Unit (1980) estimates that "80 percent of timber-generated fuel is wasted" in tobacco curing. Depletion of forest resources and any desertification related to tobacco (or any other crop) production should routinely be factored into cost/benefit analysis.

On a global scale, the fuelwood requirements for producing flue-cured tobacco contribute substantially to the serious and growing problem of deforestation in developing countries. Of the estimated 5.66 million tons of tobacco produced each year, at least 2.5 million tons are flue-cured, using fuelwood (Figure 8.4). About 55 cubic meters of stacked wood are needed to cure each ton of flue-cured tobacco, and some 87.5 million cubic meters per year of roundwood are harvested for this purpose. Since most wood for tobacco curing is harvested from the wild (rather than from sustained-yield fuelwood plantations), it is reasonable to conclude that worldwide, the equivalent of some 1.2 million hectares of open forest each year is stripped of wood for tobacco curing. WHO estimates that 12 percent of the trees cut down each year worldwide are used for tobacco curing (Frankel, 1983). This harvest is far in excess of rates of natural forest regrowth. For every 300 cigarettes made in the developing world, one tree is burned (WHO, 1980a; Muller, 1978).

In 1977, the Director of the United Nations Environmental Program warned that the shortage of firewood was rapidly becoming the poor person's energy crisis. Deforestation in the more arid areas where tobacco is frequently cultivated contributes to the massive problem of accelerating desertification in Africa. The use of wood for curing competes with its use for cooking, warmth, and construction. The relationships among tobacco curing, fuelwood shortages, and deforestation and other forms of environmental degradation are becoming increasingly clear (cf. Skutsch, 1983).

Figure 8.4: Deforestation for Tobacco Curing
(Very approximate data only.)

Number	Item	Estimate Units
1.	Annual World Tobacco Production	5.66 million tons
2.	Annual World Flue-cured Tobacco Production, Using Wood	2.50 million tons
3.	Stacked Wood Required to Cure One Ton of Tobacco (rough average) 1/	55 m³/ton
4.	Stacking Factor (amount of roundwood needed to obtain one unit of stacked wood)	0.6
5.	Annual Roundwood Harvested for Tobacco Curing (#2 X #3 X #4)	82.5 million m³
6.	Roundwood Growing Stock in Open Forest and Woodland 2/	70 m³/ha
7.	Area of Open Forest or Woodland Consumed Annually (#5 divided by #6)	1.2 million ha
8.	Hypothetical Sustained-yield Fuelwood Plantations Needed to Provide #5 3/	8.5 million ha
9.	Direct Cost of Hypothetical Plantation Needs	US$7.2 billion 4/

Notes on Figure 8.4:

1/ There is considerable regional variation in this figure; estimates vary from 70 m³/ton in Tanzania to 20 m³/ton in the Philippines.

2/ These are the environments in which tobacco is most commonly grown. As the annual increment is only about 2 m³, deforestation results.

3/ Intensively managed fast-growing fuelwood plantations can produce about 10 m³/ha/year of roundwood on a sustained-yield basis.

4/ This figure is based on a direct commercial plantation establishment cost of US$875/ha. It does not include maintenance on other recurrent costs. In the event that the land and labor needed for the plantation are not purchased, but supplied directly by the farmer, the direct (not opportunity) costs are lower (e.g., US$250/ha).

The environmental impact of tobacco production has been clearly outlined in Tanzania (Boesen and Mohele, 1979). The problem on a regional and national level is that the tobacco industry is one of the main exploiters of national forests. Depletion of forest reserves has been most advanced in the traditional tobacco areas, such as Tabora. The distance between curing barns and tree stands becomes farther each year, increasing time and effort spent on wood hauling. Tobacco cultivation is concentrated in regions where the typical vegetation is open woodland, which, after clearing, takes 30 to 50 years to regenerate. Clear felling exposes the soil to torrential rains and insolation, leading to loss of organic matter, destruction of soil texture, and consequent erosion if no counter measures are taken.

Malawi, already vulnerably dependent on international tobacco price fluctuations, may have its general economic development constrained by shortages of fuelwood. Malawi's fuelwood consumption has long and greatly exceeded production, resulting in continuing forest depletion at an annual rate of 3.7 percent (UNEP, 1982). Meanwhile, fuelwood price rises are making wood-cured tobacco less profitable. An optimistic annual woodlot planting target of 50 ha per village would barely suffice to offset the amounts of fuelwood cut for tobacco curing. In Malawi, firewood for tobacco curing on private farms is sometimes taken from communal tribal land. Cutting and burning of virtually all wooded areas in Malawi within economic range of tobacco curing is expected within eight years.

Equity and Economic Vulnerability: Tobacco proponents, aware of tobacco's health impacts and perhaps also of the role of tobacco curing in deforestation, have justified investments in tobacco production on the ground that the crop is a source of much needed foreign exchange. However, a major UNCTAD (1978) study contradicts this position. UNCTAD concludes that countries reliant on tobacco for any substantial portion of their foreign exchange are economically vulnerable to the corporations which purchase 85-90 percent of the tobacco produced in the developing world. In both developing and developed countries, 7 large tobacco corporations together effectively control every stage from leaf production to the manufacture and distribution of tobacco products worldwide. Consequently, the price of tobacco is determined largely by the fiat of administered prices, rather than the forces of a freely competitive market.

In such an "economic framework of collusion among oligopolists", UNCTAD (1978) concludes that developing countries are marginally involved in the marketing decision process. Developing countries supply 55 percent of world leaf tobacco through foreign oligopoly-controlled marketing channels; their own processed exports are minimal; they have little influence in the design, output and innovation of machinery; their aggregate revenues from tobacco are based almost exclusively on marketing decisions determined in the short, medium, and long run by the multinational tobacco corporations.

Alternative Crops: Clearly these will vary greatly by region. As oil prices continue to rise, synthetic fibers may become less competitive against natural fibers. Cotton may therefore become more economic in India and Greece, for example. Soya brings a higher return than tobacco in Brazil and the Dominican Republic. Maize competes with tobacco in Zambia, groundnuts and maize in Tanzania, and fruit and grains in Turkey. Increasing labor costs and a rural exodus may make less labor-intensive crops more competitive. Increases in fertilizer and biocide prices may reduce tobacco yields in favor of less demanding crops. People concerned with the damaging effects of tobacco will concentrate efforts on development of regionally appropriate ways to diversify crops in order to supplant tobacco.

Mitigating Deforestation Costs: Because flue curing has a high wood requirement and wood grows relatively slowly in the drier regions favored by tobacco, it is usually not economically feasible to harvest wood from natural forests on a sustainable basis for tobacco curing. Each ton of wood-cured tobacco requires roughly 33 cubic meters of roundwood--nearly one half the total growing stock on a typical hectare of tropical woodland. If only the annual natural wood increment is harvested (and the natural capital of the forest is thereby not "liquidated" on a one-shot basis), only about 2 cubic meters of roundwood could be obtained annually from each hectare. This would necessitate transporting wood very long distances to support a tobacco industry. Because of high transport costs, this is not done; instead, the woodlands relatively close to the tobacco flues are systematically stripped of most of their trees.

To prevent such deforestation and the serious environmental and social ills that result, it is essential either to phase out wood-cured tobacco production or to establish intensively managed fuelwood plantations of Eucalyptus or other fast-growing tree species. Since fuelwood plantations can produce about 10 cubic meters of roundwood per hectare on a sustained-yield basis, to produce the 82.5 million cubic meters per year of roundwood needed for tobacco curing would require establishing 8.25 million hectares of fuelwood plantations, at an estimated direct establishment cost of US$7.2 billion (plus additional social opportunity costs of alternative land uses). Thus, establishing fuelwood plantations is a viable, but expensive, alternative to phasing out wood-cured tobacco.

Since it is counterproductive to promote unsustainable forms of development (e.g., tobacco that relies upon unsustainable harvests of wood from natural woodlands), wood-cured tobacco projects should routinely incorporate fuelwood plantation components of sufficient size to meet all of the projects' fuelwood demands. Moreover, the costs of such fuelwood components need to be factored into the cost-benefit analysis of tobacco projects. Unfortunately, efforts to attach fuelwood components to tobacco projects have not always been successful because of frequent loss of fuelwood stands to fire and inadequate incentives for villagers.

Alternatively, the use of tobacco varieties capable of being air- or sun-dried would greatly reduce the problem. Solar convection curing systems and simple solar dryers also would help. Moreover, if more wood-conserving flue-curing systems were widely used, they might cut the wood demand for tobacco curing by one half.

Options to Discourage Tobacco Consumption: A variety of policy options could decrease consumption, including:

1. Education in households, in primary schools, and pre-natal clinics about the health hazards of cigarettes (partly because children of smokers are more likely to smoke than their non-smoking parent cohorts).

2. Anti-smoking publicity campaigns.

3. Warning of health risks.

4. Bans on advertising and promotion.

5. Restriction of smoking in public places.

6. Assistance programs to quit smoking.

7. Discouraging cigarette smoking more than pipe and cigar smoking (which are somewhat less harmful).

8. Preferential insurance rates commensurate with increased health and fire-related costs.

9. Increased taxation commensurate with decreased worker productivity, shortened taxable life, and increased health costs.

10. Decreasing tar levels, improving filtration; labeling content (carbon monoxide, tar, and nicotine) and roughly equating these levels with health risks. (Nicotine-impregnated synthetic tobaccos are being developed.) Legislative options regarding smoking are amplified by Roemer (1982).

As tobacco production cannot be recommended on human health, social, or environmental grounds, investment in tobacco production is ethically questionable. Furthermore, the economic justification for tobacco investments is greatly weakened by expanded cost-benefit analysis that includes such social costs as: health care costs; premature deaths and suffering from smoking-related diseases; impaired worker productivity; damage from smoking-related fires; the opportunity costs of reduced food production and consumption of the poor; harm and irritation to passive smokers (especially infants' mental development); soil degradation; environmental and health risks of biocide use; and deforestation and fuelwood depletion, among others.

References

Akehurst, B. C. 1981. Tobacco. New York, Longmans Green, 764 p.

Anderson, G.M. 1981. The tobacco industry and the third world. America 145 (Dec. 12): 374-377.

American Cancer Society, 1980. Cancer facts and figures: 1981. New York, ACS, 31 p.

Boesen, J. and Mohele, A.T. 1979. The "success" story of peasant tobacco production in Tanzania. Uppsala, Scandinavian Inst. African Studies, 156 p.

Cooper, R. 1982. Smoking in the Soviet Union. British Medical Journal 285: 549-551.

Corchoran, S. 1980. Big business in China: Sinoforeign rivalry in the cigarette industry, 1930-1980. Cambridge, Mass., Harvard University Press, 332 p.

Doron, G. 1979. The smoking paradox: public regulation in the cigarette industry. Cambridge, Mass., Abt Assocs., 141 p.

Dutta-Choudhuri, R., et al. 1959. Cancer of the larynx. J. Indian Medical Association 32(9):352-362.

Eckholm, E. 1978. Cutting tobacco's toll. Washington, D.C., Worldwatch Paper 18:40 p.

EIU, 1980. Leaf tobacco: its contribution to the economic and social development of the third world. London, The Economist Intelligence Unit, Commodity Monograph, 366 p.

FAO, 1981. Production yearbook. Rome, FAO Statistics Series 34:296 p.

Frankel, G. 1983. Tobacco is no hazard to health of Zimbabwe's economy. Washington Post (USA), September 1, 1983.

Friedman, K.M. 1975. Public policy and the smoking health controversy: a comparative study. Lexington, Mass., Lexington Books, 216 p.

Gelfand, M. et al. 1968. Carcinoma of the bronchus and the smoking habit in Rhodesian Africans. Brit. Med. J. 3(5616):468-469.

Hedrick, J.L. 1969. Smoking, tobacco and health. Washington, D.C. U.S. Health Services and Mental Health Admin. 134 p.

Hedrick, J. L. 1972. Chart book on smoking, tobacco and health. Washington, D.C., Supt. Documents, U.S. Govt. Print. Office, 46 p.

Joly, J.D. 1975. Cigarette smoking in Latin America: a survey of eight cities. PAHO Bull. 9(4):329-344.

Kark, J. D. et al. 1982. Smoking as a risk factor for epidemic A influenza in young men. New England J. Med. 307:1042-1046.

Kaufman, D. W. et al. 1983. Nicotine and carbon monoxide content of cigarette smoke and the risk of myocardial infarction in young men, New Engl. J. Med. 308:409-413.

Madeley, J. 1982. Kenyan farmers risk their lives for smokers. New Scientist (April 8): p. 67.

Madeley, J. 1983. Tobacco - the land despoiler. United Nations Development Forum (June/July): p. 12.

Muller, M. 1978. Tobacco and the third world: tomorrow's epidemic? London, War on Want, 112 p.

NEJM, 1982. Correspondence: Radioactivity in cigarette smoke. New England Journal of Medicine 307(5):309-313.

Office on Smoking and Health, 1979. Smoking and health: A report of the Surgeon General. Washington, D.C., Public Health Service, 1250 p.

Office on Smoking and Health, 1981. The health consequences of smoking: the changing cigarette. Washington, D.C., Public Health Service, 252 p.

Office on Smoking and Health, 1982. The health consequences of smoking: cancer. Washington, D.C., Public Health Service, 302 p.

OTA, 1981. Assessment of technologies for determining cancer risks from the environment. Washington, D.C., U.S. Congress Office of Technology Assessment, 240 p.

Roemer, R. 1982. Legislative action to combat the world smoking epidemic. Geneva, World Health Organization, 131 p.

Schramm, G. and Munasinghe, M. 1981. Interrelationships in energy planning: the case of the tobacco-curing industry in Thailand. Energy Systems and Policy 5(2):117-139.

Scrimgeour,E. M. and Jolley, D. 1983. Trends of tobacco consumption and incidence of associated neoplasms in Papua New Guinea. Brit. Med. Journal (30 April): 1414-1416.

Shephard, R.J. 1982. The risks of passive smoking. London, Croom Helm, 195 p.

Shepherd, A. 1975. Prospects for unmanufactured tobacco to 1984. London, The Economist Intelligence Unit No. 29:76 p.

Skutsch, M. 1983. Why people don't plant trees: the socioeconomic impact of existing woodfuel programs, Tanzania. Washington, D. C., Resources for the Future, paper D, 73 p.

Socol, M.L., Manning, F.A., Murata, Y., and Druzin, M.L. 1982. Maternal smoking causes fatal hypoxia: experimental evidence. American Journal Obstetrics Gynecology 142(2):214-218.

Sutnick, A.I. and Gunawan, S. 1982. Cancer in Indonesia. Journal American Medical Association 247(2): 3087-3088.

Taha, A. and Ball, K. 1980. Smoking and Africa: the coming epidemic. British Medical Journal 280(6219):991-993.

UNCTAD, 1978. Marketing and distribution of tobacco. Geneva, Switzerland, TD/B/C.1/205: 166 p.

UNEP, 1982. Global assessment of tropical forest resources. Nairobi, United Nations Environment Program, GEMS Info. Series 3:14 p.

USDA, 1982. Tobacco: outlook and situation. Washington, D.C. USDA Research Service (March):33 p.

USDA, 1983. Tobacco: outlook and situation. Washington, D.C. USDA Research Service (March):30 p.

Van Wambeke, A. 1975. Management properties of oxisols in savanna ecosystems. (364-371) in Bornemisza, E. and Alvarado, A. (eds.). Soil Management in Tropical America. Raleigh, N.C., North Carolina State Univ:565 p.

WHO, 1975. Smoking and its effects on health. Geneva, World Health Organization, Tech. Rept. 568:100 p.

WHO, 1979. Controlling the smoking epidemic. Geneva, WHO, Expert Ctee on Smoking Control, Tec. Rept. 636:87 p.

WHO, 1980a. Save the rainforests. Bull. IUCN 11(5):p. 30.

WHO, 1980b. Smoking or health: the choice is yours. Geneva, World Health Organization, 39 p.

World Development Report, 1981. Washington, D.C., World Bank: 192 p.

9
Cotton

Cotton (Gossypium), the world's major fiber crop, is likely to increase in importance as petrochemicals for the manufacture of synthetic fibers become less affordable and as the population grows. Between 1950 and 1960, cotton cultivation expanded enormously, mainly in the arid and semi-arid regions of the Middle East and Latin America. It is now the most important of all non-food crops, and covers about 5 percent of the world's cultivated land area. Between 1948 and 1967, cotton production grew from 650,000 to 1,350,000 tons in the Middle East, and from 775,000 to 1,450,000 tons in Latin America. World cotton trade averaged US$7 billion a year between 1978 and 1980.

While cotton has become a major source of much-needed foreign exchange in many developing countries, the massive increases in cotton production were achieved at profound social and environmental costs. Land use conversion from foodstuff to cash crop production has, in some cases, exacerbated rural poverty, and the seasonal nature of cotton cultivation has created periods of rural underemployment and unemployment in the intervening periods. The cotton plantation's requirement for large numbers of seasonal migrant laborers, who are often inadequately housed, greatly increases local public health problems. Because cotton yields a comparatively high financial return, it is one of the few crops which justifies the high investment costs of major irrigation schemes. Many developing countries, often with bilateral or multilateral assistance, have irrigated vast stretches of land for cotton cultivation. The negative environmental impact of such development is significant unless projects are carefully designed and monitored.

To obtain maximum returns, developing countries have sought, and in some cases achieved cotton yields of over 1,000 kg/ha (over twice the average cotton yield in the United States). However, under intensive cropping on large areas, the risk of pest buildup is high. Moreover, such levels of production have been sustained only by means of increasing biocide use. Both the immediate and long-term environmental and health damage wrought by excessive use of agro-chemicals is a major and intensifying concern associated with cotton production.

Cotton can also be grown by individual farmers within
small-scale, traditional systems which place less strain on the
environment. Moreover, cottonseed presscake from small-scale cotton
production could supplement protein levels in local diets. "The
presscakes ... of soybeans, cottonseed, peanuts, and sesame seeds
are perhaps the most accessible untapped source of protein for human
consumption; today much of it is wasted" according to Ehrlich and
Ehrlich, as recently as 1976. Other cotton byproducts have a wider
range of uses (Lewis and Richmond, 1972).

Water Management: Two major concerns associated with
cotton (and all other irrigated crops) are the salination and
waterlogging of soil by excessive irrigation and poor drainage; and
the spread of water-related diseases are consequences of increased
vector breeding opportunities. The negative impacts of irrigation
and measures to mitigate them are outlined in Chapter 20.
Environmental problems associated with large irrigated cotton
schemes are discussed by Giglioli (1979). In addition to the
improved habitat for malaria-transmitting Anopheles mosquitoes that
irrigation can provide, the heavy biocide spraying associated with
cotton cultivation can produce biocide resistance in these
mosquitoes, making malaria control increasingly difficult and in
some cases impossible. Recent outbreaks of malaria in regions
hitherto free of the disease have been attributed to induced
resistance in mosquitoes to biocides used in cotton production.

On well-drained heavily fertilized soils, large quantities
of nutrients (especially nitrates) leach downward and enter local
water bodies. Assuming a soil absorption rate of 50 percent for the
applied nitrogen (N) and an application rate of 220 kg N/ha (Hearn,
1975), perhaps as much as 100 kg of nitrogen is lost into the
drainage from each hectare. Similarly high levels of nutrient
runoff into watercourses have resulted in eutrophication and, when
combined with biocide residues, the destruction of fisheries.

In contrast, cotton can be successfully grown using
traditional methods under rainfed conditions with only about 1 meter
of rain per crop. Yields are low, averaging 300 kg/ha in Northern
Nigeria (Kowal & Faulkner, 1975). However, since little fertilizer
and biocide is used, there is little pollution of the drainage water
or human health hazard.

Soil Management: The major soil impacts of cotton
cultivation are associated with irrigation, mentioned above, and
soil erosion. A deep-rooting crop, cotton thrives best on deep,
fertile soils. Good drainage is particularly important for
irrigated cotton even though the crop is fairly tolerant of salinity.

As cotton exhausts soil nutrients quickly, continuous
cotton cultivation rapidly depletes soil fertility. Although cotton
fiber contains very small amounts of minerals, and cottonseed oil
practically none at all, the seed itself contains large amounts of
minerals. On good soils a high-yielding crop takes up about 160 kg

N/ha. Nitrogen must therefore be replaced, along with other nutrients, in fertilizer supplements. Where phosphate fertilizer is also applied, significant quantities of phosphates may be carried by runoff into lakes and streams, thus encouraging eutrophication. If the cotton is intercropped with a legume, the need for nitrogenous fertilizers can be reduced.

Prompt shredding and plowing under of all leaves, stems, roots, and other harvest residues reduces the survival of pests and diseases. Use of residues as a fuel, though attractive as an energy supplement, can present a hazard as pests survive in the stored residues prior to burning. Burning the residues volatilizes nitrogen, precluding its return to the soil and thus accelerating soil impoverishment, though some benefits can be obtained if ash is returned to the soil. Where the cotton is grown under traditional, non-intensive systems, less care is usually taken to destroy harvest residues, with the result that the net nutrient loss is less.

Pest and Weed Management: More biocides are used in cotton production than in any other crop. Cotton cultivation receives by far the bulk of biocides used in developing countries. Even in the United States, cotton absorbs 47 percent of all insecticides used. The frequent and heavy application of biocides on intensively cultivated cotton and the resulting serious major health and environmental effects are well described in the literature (Giglioli, 1979; FAO, 1973; ICAITI, 1977). Since the early 1970s, of the many Central Americans who have been poisoned by biocides, hundreds have died and thousands more suffer from sub-clinical intoxication. As noted above, malaria vectors in cotton-growing areas have developed resistance to chemical control. Biocidal contamination of beef has cost Guatemala and Nicaragua millions of dollars in foreign exchange, because of inability to meet the biocide residue standards of importing nations.

FAO divides cotton production and associated crop protection methods into a series of distinct phases. In the subsistence phase, the crop is usually grown without irrigation, enjoys no organized pest protection, and yields less than 225 kg of lint per hectare. Protection against pests is derived from natural control, plant resistance, infrequent insecticide treatments, hand-picking of pests, traditional cultural practices, and the usual absence of large contiguous areas under the crop. In the exploitation phase, cotton is usually irrigated and chemical pest control prevails. High yields of seed and lint are achieved. In the crisis phase, attempts are made to gain effective control through more frequent applications of biocides that are started earlier in the crop season and continued later into the harvest period. After treatment, pest populations tend to escalate to higher levels, having become so tolerant of the biocide that treatment ultimately proves economically useless. Alternative biocides are then substituted, and these in turn also become useless as pest populations acquire resistance. New pests intrude as their natural predators are unwittingly eliminated by biocides; they become serious and consistent cotton ravagers. In the disaster phase, costs of biocide use reach the point where cotton no longer can be grown profitably, and lands are removed from cotton production. In the fifth stage - the integrated control phase - a crop production system is structured using a variety of control procedures, rather than relying only on chemical biocides. Fullest use is made of cultural and biological methods of pest control (FAO, 1973). Ideally, this should be the first and only stage in judiciously designed projects. Integrated pest management (IPM) systems for cotton are improving. For example, safer substances such as the bacteria Bacillus thuringiensis and synthetic pyrethrin-based biocides are increasingly being used, as in Tanzania's Geita cotton project and Madagascar's cotton project (Chapter 17).

Public Health Concerns

Uneven employment patterns and the need for large numbers of unskilled, low-cost laborers for harvesting generate an annual migration of several million people a year. Entire families migrate to cotton-growing areas where frequently they are stricken with numerous illnesses related to unhealthy living conditions. As a result, all such sudden population influxes require stringent public health and sanitation precautions. Water-related diseases and maladies brought in by migrant workers are prime concerns; medical screening of the migrant families and treatment of disease carriers can prevent such problems. Adequate and safe water and provision for waste disposal are essential.

Mechanical harvesting of cotton is seldom a satisfactory alternative to hand labor, despite the public health and social dislocation risks and (sometimes) increased labor costs. The machinery is complicated, expensive to purchase, maintain, and operate, and is used only briefly each year. Furthermore, agricultural mechanization of any kind runs the risk of aggravating rural unemployment or underemployment, thereby further increasing poverty and sometimes urban migration, or migration into forests or other lands unsuited for agricultural settlement (see Chapter 21).

References

Adkisson, P.L. et al. 1982. Controlling cotton insect pests: a new system. Science 216:19-22.

Brown, H.B. and Ware, J.O. 1958. Cotton. New York, McGraw-Hill, 566 p.

Dema, P. et al. 1974. Integrated control of cotton pests in Thailand. Bangkok, Thailand. Ministry of Agriculture. Plant Protection Service Technical Bulletin 32: 27 p.

Ehrlich, P.R., Erlich, A.H. and Holdren, J.P. 1977. Ecoscience: population, resources and environment. San Francisco, Freeman, 1,051 p.

Falcon, L. and Smith, R. 1973. Guidelines for integrated control of cotton insect pests. Rome, FAO: 92 p.

Giglioli, M.E.C. 1979. Irrigation, anophelism and malaria in Adana, Turkey. Washington, World Bank, 86 p.

Hagan, R., Haise, H. and Edminster, T. (eds.) 1967. Irrigation of agricultural lands. Madison, Wisconsin. American Society of Agronomy, 1,180 p.

Hearn, A.B. 1975. Ord valley cotton crop: development of a technology. Cotton Growers Rev. 52:77-102.

ICAITI. 1977. An environmental and economic study of the consequences of pesticide use in Central American cotton production. Guatemala City, ICAITI/UNEP:273 p.

Kowal, J.M. and Faulkner, R.C. 1975. Cotton production in northern states of Nigeria in relation to water availability and crop water use. Cotton Growers Rev. 52:11-29.

Lewis, C.F. and Richmond, T.R. 1972. Cotton improvement (4-14) in Cotton. Ciba-Geigy Agrochemicals Tech. Monogr. 3:30 p.

Malima, K.A. 1971. Cotton, agricultural and economic transformation in Tanzania. University of Dar es Salaam, 16 p.

Pentice, A.N. 1972. Cotton with special reference to Africa. London, Longmans Green, 282 p.

Seminar on cotton production research. 1966. Cotton breeding for disease resistance and the control of diseases. Lima, International Cotton Advisory Committee, 75 p.

UNDP, 1978. Cotton: new initiatives from seed to sale. New York, UNDP/TCDC Case Study 8:12 p.

Verschaege, L. 1972. Problems of cotton growing in developing countries (15-22) in Cotton. Ciba-Geigy Agrochemicals Tech. Monogr. 3:80 p.

World Bank, 1979. Occupational safety and health guideline: cotton ginning. Washington, D.C., World Bank, Office of Environmental Affairs, 3 p.

10
Oil Palm

"Our money grows on trees".
--Malaysian T-shirt stencil

Oil palms (**Elaeis guineensis**) produce the highest oil yields of any crop (averaging 1.25-1.50 tons/acre/yr over the life of the crop). Palm oil has an increasing range of uses, both as a foodstuff (because it is low in unsaturated fatty acids) and in industry. Residual kernel cake and, to a lesser extent, press residue is valuable as feed for large and small livestock. Since it is a perennial like rubber, oil palm has soil conservation value because the harvest does not kill the tree, as is the case with many other crops. The major environmental concerns associated with oil palm are deforestation resulting from the expansion of plantations and water pollution from oil palm factory effluents.

Land and Water Conservation: Oil palms, generally planted at a density of 140-160 per hectare and spaced about nine meters apart in a triangular array, protect the soil from erosion. Although the canopy it forms is not as closed and therefore not as protective as that of rubber, the ground layer is usually more protective. Oil palm is preferably grown on flat or gently undulating land to facilitate harvest of the heavy fruit bunches, but its perennial and protective nature--augmented by associated cover crops--makes it an environmentally preferred crop on steeper slopes. Cover crops control erosion in oil palm plantations regardless of the slope. Where land is steep or where rainfall is exceptionally heavy, terraces or platforms can be cut into the slope to facilitate harvest and further reduce erosion.

At reduced planting densities, oil palm intercropping with bananas, coffee, cacao and other tree crops also controls erosion and reduces economic dependence on oil palm monoculture. However, where a processing factory is necessary for modern productive systems, intercropping is limited by the demand for dense palm plantings.

In West Africa, where the oil palm is an indigenous food crop, it is grown by the villagers as one component of a compound farm or home garden comprising a wide variety of annual and perennial food crops. Food cropping can be carried out successfully within a plantation, provided that adequate fertilizer inputs are applied to compensate for the large nutrient depletion from the crop mix. This possibility merits consideration in any scheme where local farmers traditionally tend to inter-crop.

Control of soil erosion depends on the maintenance of a ground cover of legumes, vines or grasses. These not only reduce soil erosion (particularly on sloping land), but also help to conserve soil organic matter improve soil structure and aeration (as in the case of rubber, Chapter 11). Legumes also contribute significant amounts of phosphate (P), potassium (K), calcium (Ca) and magnesium (Mg) as well as nitrogen (N) to the palm. By increasing infiltration on heavy soils, water may be better conserved, although water uptake by ground covers and intercrops is significant (Chapter 29).

Palm oil (a hydrocarbon) itself is oligotrophic, containing insignificant amounts of minerals, but the fruit bunches are eutrophic, and contain large quantities of minerals which must be replaced if productivity is to be sustained. Replenishment, either by chemical fertilizer or by nutrient recycling (including ash from fiber boiler fuel) is therefore, necessary. Since fertilizer costs can amount to 50 percent of field production costs, useful economies can be achieved by recycling minerals remaining in the fruit bunches after the oil has been extracted. Press cake and sludge have been successfully used for this purpose in some projects. Recycled fruit bunch ash and mill effluent reduce waste, decrease chemical fertilizer requirements, conserve water, and minimize the pollution caused by discharging effluent into small streams.

While leguminous and other nitrogen-fixing plants can and, in well-designed projects, do supply oil palms with most of their nitrogen needs, yields that are high enough to repay major a commercial investment can be obtained only with substantial N, K, P, and Mg applications. Since manuring can amount to 50 percent of the annual agricultural expenditures of an oil palm estate using exogenous sources of fertilizer, the use of local estate manure will reduce costs.

Deforestation: A major environmental cost of modern oil palm projects is the loss of biologically diverse tropical forest, which is cleared for establishment of plantations. The loss of unique genetic material and the loss of habitat for animals and plants and--most importantly--for indigenous peoples, is outlined by the World Bank (1982) and in Chapter 12. The effects of deforestation can be palliated by careful site selection.

Pest and Weed Management: Since oil palms are not as naturally susceptible to insect and other pests as many other crops, there is little need for biocides. The chemical control of plant diseases and insect attacks can be minimized by observing sound agronomic practices, such as removal of diseased organs and individual trees to prevent disease spread.

Non-chemical controls, including integrated pest management (IPM), reduce enviromental risks. Paradoxically, outbreaks of seven major oil palm pests in Malaysia were shown to be directly related to the annihilation of natural enemies through indiscriminate use of biocides. The effectiveness of natural enemies is increased with less extensive spraying and selective weeding, such as avoiding removal of weeds that provide food and cover for the natural enemies (Chapters 22 and 23).

In those countries where oil palm was introduced without a pollinator (e.g., in Malaysia in 1917), the major costs of pollinating by hand can be avoided by introduction of the West African weevil Elaeidobius kamerunicus. Following its large-scale release in Malaysia in 1981, this weevil raised fruit set in the first subsequent harvest from 48 to 76 percent (Syed, et al., 1982), increasing Malaysian palm fruit yield by 130,000 tons, thereby earning US$57 million annually in foreign exchange. This introduction makes integrated pest management especially important if the pollinating insect is to be unhindered.

In oil palm plantations, as in rubber, leguminous ground covers reduce weed invasion. Any unavoidable use of herbicides should be applied only by adequately trained personnel, with supervision, monitoring and all appropriate safety precautions observed (Chapter 22).

Effluent Disposal: The massive expansion of oil palm production over recent years has created the problem of disposing of the large amounts of effluent, mainly sterilizer condensates and clarification sludge, produced during the extraction of oil. Two to three tons of aqueous liquid effluent are produced for every ton of finished oil. For every ton of fresh fruit bunches (FFB) processed, about 600 liters of effluent are produced. Disposal is exacerbated by the particularly high biochemical oxygen demand (usually at least 20,000 mg per liter) of crude effluent. This is around 100 times higher than the figure for raw sewage. In major producing countries, the total pollution load from commercial palm oil production approaches that of the sewage from the country's entire population. One palm oil factory handling 20 tons FFB/hour pollutes as much as a city of 200,000 people (Petitpierre, 1982). Fortunately, however, the effluent is essentially free from actively toxic constituents.

Despite its harmful effects on water quality (harming fish populations and impairing many other water uses), oil palm effluent can be valuable as a source of soil nutrients and renewable energy. Consequently, increasing efforts are being made to transform the problem of effluent disposal into an asset.

There are many effective methods to reduce the environment damage of oil palm effluent. Microbial decomposition in aerobic or anaerobic ponds, for example, purifies the effluent into a useful, nutrient-rich irrigant for the plantations. Also, the pond solids that are allowed to dry out are periodically scraped off for use as fertilizer on plantations. These systems, practised increasingly in Malaysia, are inexpensive, but require much space.

Oil palm effluent consists of 10 percent crude protein, 12 percent crude fiber, 20 percent fatty materials, 11 percent ash, and 47 percent nitrogen-free extract (mainly starch, sugars and other carbohydrates, and lignin). It is therefore useful as a constituent of livestock feed, or as a carbohydrate and nitrogen source for the manufacture of single-celled protein. Oil palm sludge-based alcohol and biogas production is gaining favor in Malaysia. About 5 liters of fuel oil equivalent is produced as biogas from each ton of fresh fruit bunches (250,000 liters/year or US$62,500 at US$0.25/liter in the 20 tons/hour factory mentioned).

Palm Oil Refining: The refining of palm oil creates further environmental problems (Watson and Meierhoefer, 1976), such as large volumes of oily water from the scrubbers, foul smells from the deodorization stage, and the need to dispose of such wastes as nickel-contaminated diatomaceous earth from the dehydrogenation process, and spent clay from the bleaching process. The oil-water separation is preferably carried out using a device approaching the American Petroleum Institute design. The effect of the odors can be minimized by proper plant siting. The nickel catalyst and the heavy burdens of oil (up to 40 percent) in the diatomaceous earth and clay can be reclaimed. The dumping of oil-rich earth and clay can lead to spontaneous combustion and contamination of drainage water unless carried out with care (Svensson, 1976).

Animal Damage: Elephants, porcupines, rats and other rodents, deer, pigs and many other animals can kill or damage economically significant amounts of oil palms. In one major oil palm project in Malaysia, elephants caused damage of many millions of dollars. Faced with loss of their natural forest habitat, the elephants invaded and destroyed many young oil palms, apparently on purpose, rather than for food, or accidentally. These losses could have largely been prevented at low cost by securing enough natural habitat for the elephants in national parks and other protected areas. Additional information on managing elephants in oil palm and other agricultual projects is provided by Seidensticker (1983).

References

Agamuthu, P., Furtado, J. I. and Broughton, W.J. 1980.
Establishment and subsequent effects of legume covercrops on the
development of young oil palms. (561-568) in Furtado, J.I. (ed.)
Tropical Ecology and Development. Kuala Lumpur, Int. Soc. Trop.
Ecol. 1,383 p.

Aderungboye, F.O. 1977. Diseases of the oil palm. PANS V.
23(3):305-326.

Chan, K.C., Goh, S.H. and Tan Wang Ing. 1976. Utilization of oil
palm nut shells. Kuala Lumpur, Planter 52:127-130.

Chin, K.K. 1978. Palm oil waste treatment by aerobic process.
(505-511) in Water pollution control in developing countries.
Bangkok, Proc. Inter. Conf., 2 vols.

Collier, H.H. and Chick, W.H. 1977. Problems and potential in the
treatment of rubber-factory and palm oil mill effluents.
Planter 53:439-448.

Corley, R.H.V., Hardon, J.J. and Wood, B.J. (eds.) 1976. Oil palm
research. Amsterdam, Elsevier Scientific Publishing, 532 p.

Dalzell, R. 1977. A case study on utilization of effluent and
by-products of palm oil production by cattle and buffaloes on a
oil palm estate. Cited by Wood, 1977.

Davis, Y.B. 1978. Palm oil mill effluent: A review of methods
proposed for its treatment. Trop. Sci. 20(4):233-262.

Devendra, C. 1978. The utilization of feeding stuff from the oil
palm plant. (116-131) in Devendra, C. and Hutagalung, R.I.
(eds.) Proc. Symp. Feeding-stuff for Livestock in S.E. Asia.
Kuala Lumpur, Malaysian Society of Animal Production, 400 p.

Devendra, C. and Muthurajah, R.N. 1977. The utilization of oil
palm by-products by sheep. (103-123) in Earp, D.A. and Newal, W.
(eds.) International Developments in Oil Palm. Kuala Lumpur,
Incorporated Society of Planters. 545 p.

Eapen, P.I. 1977. Palm oil mill effluent as a source of water and
nutrients for plants. FAO Preprint (OC/77/17). Benin City,
Nigerian Institute for Oil Palm Research, (v.p.).

Genty, P., Desmier de Chenon, R., Morin, J.P., Dorythkowski, C.A.
1978. Oil palm pests in Latin America. Oleagineaux
33(7):325-419.

- 76 -

Godin, V.J. and Spensley, P.C. 1971. Oils and oil seeds. London,
Tropical Products Inst., Crop & Products Digest 1: 170 p.

Hartley, C.W.S. 1977. The oil palm. London, Longmans Green, 806 p.

Hemming, J.L. 1977. The treatment of effluents from the production
of palm oil. (79-101) in International developments in palm
oil. Kuala Lumpur, Incorporated Society of Planters. 545 p.

Hutagalung, R.I., Chang, C.C., Syed Jalaludin and Webb, B.H. 1975.
The value of processed oil palm sludge as a feed for chicks.
Malay. Agric. Res., 4:53-60.

Hutagalung, R.I., Chang, C.C., Toh, K.M. and Chan, H.C. 1977.
Potential of palm oil mill effluent for growing/finishing pigs.
Planter 523:2-9.

H.V.A - International N.V.A. 1972. Long-term prospects for palm oil
on the world market for fats and oils. Amsterdam. 57 p.

Khera, H.S. 1976. The oil palm industry of Malaysia: an economic
study. Kuala Lumpur, University of Malaysia. 354 p.

Kirkaldy, J.L.R. 1976. Possible utilization of by-products from
palm oil industry. Planter 52:118-126.

Kirkaldy, J.L.R., 1979. Treatment of oil palm sludge. (243-248)
Moo-Young, M. and Farguh, G.J. (eds.) in Waste Treatment and
Utilization--Theory and Practice of Waste Management. London,
Pergamon Press, 572 p.

Lynan, J. 1972. West African production and exports prospects for
palm oil and palm kernel oil to 1980. USDA, Foreign
Agricultural Service, 36 p.

Muthurajah, R.N. and Bin, P.T. 1976. Manufacture of paper pulps from
oil palm empty bunch wastes. Kuala Lumpur, Malaysian
International Symposium Palm Oil Processing Marketing: 10 p.

Ng, S.K. 1972. The oil palm, its culture, manuring and utilization.
Berne, International Potash Institute, 142 p.

Petitpierre, G. 1982. Palm oil effluent treatment and production of
biogas. Oleagineux 37(7):372-373.

Pollak, P. 1976. Prospects for palm oil. Washington, D.C. World
Bank Commodity Paper No. 23: 18 p.

Rajagopalan, K. and Webb, B.H. 1975. Palm oil mill waste recovery
as a by-product industry, II. Planter 51:126-132.

Seidensticker, J. 1983. Elephants in agriculture and forest development projects. Washington, D.C., World Bank, Office of Environmental Affairs, 35 p.

Seow, C.M. and Wren, W.G. 1977. Studies on testing palm oil mill effluent by anaerobic digestion. Cited by Wood, 1977.

Svensson, C. 1976. Use or disposal of by-products and spent material from the vegetable oil processing industry in Europe. J. Amer. Oil Chemists Soc. 53:443-445.

Syed, R.A., Law, I. H. and Corley, R.H.V. 1982. Insect pollination of oil palm: introduction, establishment and pollinating efficiency of Elaeidobius kamerunicus in Malaysia. Kuala Lumpur, Planter 58:547-561.

Syed, R.A. and Shad, S. 1977. Some important aspects of insect pest management in oil palm estates in Sabah, Malaysia. (577-590) International Development in Oil Palm. Kuala Lumpur, Malaysia, 804 p.

Thillaimuthu, J. 1978. The environment and the palm oil industry: A new solution: incineration of sludge. Planter 54(626):228-235.

Tropical Products Institute, 1965. The oil palm. London, Ministry of Overseas Development, 167 p.

Turner, P.D. 1974. Oil palm cultivation and management. Kuala Lumpur, Incorp. Soc. Planters:672 p.

Usoro, E.J. 1974. The Nigerian oil palm industry: Government policy and export production. Ibadan University Press, 153 p.

Watson, K. S. and Meierhoefer, C.H. 1976. Use or disposal of by-products and spent material from the vegetable oil processing industry in the U.S. J. Amer. Oil Chemists Soc. 53:437-442.

Webb, B.H., Hutagalung, R.I. and Cheam, S.T. 1976. Palm oil mill waste as animal feed: processing and utilization. Malaysian International Symposium on Palm Oil Processing and Marketing. Kuala Lumpur, 21 p.

Webb, B.H., Hutagalung, R.I. and Cheam, S.T. 1977. Palm oil mill waste as animal feed: processing and utilization. (125-145) in International developments in palm oil. Kuala Lumpur, Incorporated Society of Planters, 545 p.

Williams, C.N. 1970. Oil palm cultivation in Malaysia: technical and economic aspects. Kuala Lumpur, University of Malaysia Press, 205 p.

Wood, B.J. 1978. Research in relation to natural resources - oil palm. Planter 54:414-441.

Wood, B. J. 1977. Oil Palm research. Amsterdam, Elsevier Scientific, 532 p.

World Bank 1979. Palm oil industry guidelines. Washington, D.C., World Bank, Office of Environmental Affairs, 7 p.

World Bank, 1982. Tribal peoples and economic development:human ecologic considerations. Washington, D.C., World Bank, 111 p.

11
Rubber

Natural rubber (**Hevea** **brasiliensis**), a renewable hydrocarbon, is becoming increasingly valuable as the petroleum upon which the manufacture of synthetic rubber depends increases in price and becomes depleted. Between 1967 and 1974, inflation-adjusted prices of petroleum feedstocks for synthetic rubber increased between 140 and 565 percent, while the price of crude oil itself rose 74 percent between 1978 and 1980 (Figure 11.1 and Chapter 21). Only half a ton of crude oil is needed to produce one ton of natural rubber. In contrast, 3.5 tons of crude are required to produce one ton of styrene butadiene rubber, and 5.5 tons of crude for one ton of polyisoprene (Goering, 1982).

Figure 11.1: Increases in Synthetic Rubber Feedstock and Monomer Prices, 1967-74.

| | US$ per ton | | Increase 1967-74 |
	1967	1974	(%)
Naphtha	20	130	550
Benzene	74	500	565
Butadiene	185	385	108
Styrene	180	800	344
Isoprene	275	660	140

Source: Sekhar, 1976.

Production of natural rubber also has a high natural resource conservation value in that the latex is derived from a perennial (25-30 years) tree crop. Moreover, the harvest does not kill the plant, as is the case with many other crops. Deforestation for the expansion of rubber cultivation and the pollution from rubber processing are the major environmental concerns associated with this crop. Conversion of non-forest land to rubber plantation

is environmentally preferable. If this is not feasible, then
secondary forest should be converted rather than primary forest; and
logged-over primary forest rather than virgin forest.

Land and Water Conservation: As a canopy of rubber trees
tends to simulate the natural forest canopy, well-managed rubber
cultivation is more environmentally protective than most crops,
although less effective than well-managed long rotation timber
production. Rubber latex, almost entirely a hydrocarbon, is
oligotrophic (containing only minor quantities of minerals); thus,
its harvest incurs a lower net mineral loss from the ecosystem than
most crops. Erosion and soil exhaustion under well-managed rubber
is far less likely than under annual crops, especially on slopes and
in high rainfall areas. Although rubber is grown in areas of high
rainfall and often on steep terrain where the potential for severe
erosion is high, the abundant supply of water stimulates undergrowth
vegetation which, in turn, effectively reduces the effects of
raindrop impact. Under well managed rubber plantations in which
natural undergrowth is maintained between the rows of trees, soil
erosion rates are often negligible.

In rubber cultivation, erosion can be held within tolerable
limits on slopes of up to about 18°. On 8° - 18° slopes,
terraces approximately four meters wide are necessary, while contour
bunds can adequately control erosion on slopes of 3° to 8°.
Since erosion is strongly related to rainfall, special provisions
such as silt pitting can mitigate problems in gently sloping areas
where rainfall exceeds 2.5 to 3.0 meters per year (Chapters 21 and
25).

Control of erosion by maintaining a vegetative ground cover
supplements mechanical controls. In planting cleared forest land,
natural regeneration can usually provide this cover. However,
following severe burns which may retard natural regrowth, leguminous
cover crops should be planted, such as mixtures of Pueraria
phaseoloides, Centrosema pubescens, and Calopogonium mucunoides.

Besides preventing erosion, leguminous ground covers play
an important role in fixing and maintaining nitrogen in the soil.
In young plantations, this function greatly reduces nitrogenous
fertilizer needs, which in any case are not large. Shade-tolerant
ground cover, such as Calopogonium mucunoides, assists when the
rubber canopy closes. The Rubber Research Institute of Malaysia
(RRIM) (1974) reports that in the sixth and seventh year of trials,
the total nutrient returns from C. caeruleum are sufficient to meet
the trees' total requirement for nitrogen. While spontaneous
carpets of indigenous plants protect soil from erosion, additional
nitrogen must be applied if yields are to equal those of rubber
cultivated in association with leguminous ground covers.

Intercropping with certain food crops also can protect soil
from erosion, depending on rainfall, slope, choice of the intercrop,
and other factors. However, under smallholder conditions, planting

density of about 600 trees per hectare is preferred. At this density there is always competition for nutrients. Where conditions permit, intercropping can maximize land use efficiency. Bananas, papaya, pineapples, watermelons, upland rice, maize, soybeans, mung beans, yams, ginger, and sweet potato have all been successfully grown as intercrops with young rubber where careful management avoids competition with the rubber. Multicropping and mixed farming with rubber increases food self-sufficiency and reduces over-reliance on rubber monocultures. With precautions, intercropping while replanting rubber may be possible.

Intercropping is usually uneconomical in rubber plantations established on newly-cleared forest land because of the presence of stumps, roots or logs. Legumes can be planted as a cover crop however, for soil conservation purposes. Where old rubber trees are replaced by new rubber trees, the intercropping becomes easier. Either perennial (e.g., bananas, cocoa, coffee) or annual crops may be used, but perennials are environmentally preferred. Slope must also be taken into account: annual crops should not be planted on land with greater than 16^0 slopes; between 8^0 and 16^0 slopes, annual cropping is possible with terracing; and from 0^0 to 8^0, annual cropping is feasible, but still more prone to erosion than perennial crops, even with contour ploughing.

Rubber production can also be linked to ruminant livestock production. Experiments at the RRIM (1975, 1976) with the fast growing pigeon pea, Cajanus cajan, indicate it is promising as a cover crop at certain stages because it withstands drought and grows vigorously on soils having poor structure. It also is a source of nutritious animal forage and even when six to eight cuttings were harvested during the first year, the plants continued to grow satisfactorily and the feed conversion ratio appeared favorable during the second year. Care is needed, however, because ruminants may damage the trees and, being selective grazers, may even stimulate weed growth (e.g., lalang grass, Chapter 1) by eating the legumes. Where labor is abundant, then keeping the ruminants penned or tethered and hand-cutting the feed for them reduces this problem. Other forms of livestock production also may be integrated with rubber production. Poultry rearing under rubber in Malaysia has proved profitable if undertaken as an extensive, low-cost family operation. Such diverse production systems featuring rubber, food crops and livestock, approach the environmentally ideal pekarangan or "home gardens" of Indonesia, "Kandy gardens" of Sri Lanka, and "Mayan gardens" of Central America (Chapter 12).

A major environmental cost of rubber cultivation, however, is the loss of forest converted to create these plantations, unless previously cleared or non-forest land is used. This cost is discussed in Chapters 17 and 24. Operational options to mitigate loss of such habitat and other resources are outlined in Chapter 24.

Pest and Weed Management: Possibly because of their latex, rubber trees are little affected by most insects, and thus seldom require application of biocides. Agrochemical control of plant diseases and insect attacks is minimized by removing diseased stumps and other points of infestation. Where integrated pest management has been developed for rubber, such techniques further reduce the disadvantages of biocides (Chapters 17 and 18).

The use of herbicides to control weeds in young rubber plantations can be minimized through manual removal or by cultivating leguminous or other ground crops. Where weeds cannot be controlled manually (because of labor limitations) or biologically, because of labor limitations, all precautions must be followed in herbicide use; under such circumstances, the normal practice of spot killing is less damaging than other methods (Chapter 22).

Rubber Processing: Rubber effluent discharged on a large scale into rivers will cause pollution. The large amount of organic material in the effluent is a substrate for the growth of microorganisms which create offensive odors and a high biochemical oxygen demand. The resultant deoxygenation of rivers harms fish and other biota. Although this damage currently is widespread, it can be prevented. Each small, local processing center creates minor problems in terms of effluent, while processing at large latex centers provides opportunities for control of pollution. Biological ponding is a cheap and simple precaution. In Malaysia, recovered rubber-processing wastes have been successfully used for fish culture. Ammonia, added during processing to prevent premature coagulation of the serum from the trees, is usually wasted with the effluent, but this can be recycled as a basis for algal protein production (Feachem, 1977). Simple traps with a retention time of not less than 12 hours for the total daily effluent facilitate precipitation of total suspended solids to tolerable levels, thus leaving the cleared effluent compatible for multiple water uses downstream. The trapped material can be removed and profitably recycled as fertilizer. Thus, rubber processing by-products can become a useful resource, rather than a pollutant. In any case, rubber processing is not as polluting as processing palm oil (Chapter 10).

References

Chin, P.S., Singh, M.M., John, C.K., Karim, M.Z.A., Bakti, N.A.K., Sethu, S., Yong, W.M. 1978. Effluents from natural rubber processing factories and their abatement in Malaysia. (229-242) in Water Pollution Control in Developing Countries. Proc. International Conference, Bangkok, 1,027 p. (2 vols.)

Ching, L.S. 1969. An agro-economic study of intercrops on rubber smallholdings. Malaya, Rubber Research Inst., Econ & Planning Div. Rept 6: 56 p.

Collier, H.M. and Chick, W.H. 1977. Problems and potential in the treatment of rubber-factory and palm oil mill effluents. Planter 53:439-448.

Dunham, D.M. 1974. Spatial implications in the competition between natural and synthetic products: with special reference to case of rubber. The Hague, Institute of Social Studies, 51 p.

Economist Intelligence Unit, 1974. Rubber and the energy crisis: the implications to the rubber industry in Western Europe, North America, South East Asia and Japan. London, The Economist, 33 p.

Economist Intelligence Unit, 1980. Natural rubber: a detailed examination of aspects of the international natural rubber agreements and its wider implications. London, The Economist, 375 p.

Feachem, R., McGarry, M., and Mara, D. (eds.) 1977. Water, wastes and health in hot climates. New York, Wiley, 399 p.

Grilli, E.R., et al. 1980. The world rubber economy: structure, changes and prospects. Washington D.C., World Bank, Staff Occasional Papers No. 30:204 p.

Goering, T.J. 1982. Natural rubber. Washington, D.C., World Bank, Sector Policy Paper, 68 p.

Lim, S.C. 1976. Land development schemes in peninsular Malaysia; a study of benefits and costs. Rubber Research Institute of Malaysia: 386 p.

Polhamus, L.G. 1962. Rubber: botany, production and utilization. London, L. Hill Publishers, 449 p.

Rubber Research Institute of Malaysia, 1974. Annual Report. Kuala Lumpur, Muhibbah Enterprise, 250 p.

Rubber Research Institute of Malaysia, 1974. Course of soils: soil management and nutrition of rubber. Kuala Lumpur, RRIM: 195 p.

Rubber Research Institute of Malaysia, 1975. Annual Report. Kuala Lumpur. Muhibbah Enterprises, 259 p.

Rubber Research Institute of Malaysia, 1976. Annual Report. Kuala Lumpur. Muhibbah Enterprises, 302 p.

Sekhar, B. 1976. The future of natural rubber, a forecast. Kuala Lumpur. Malaysian Rubber Research and Development Board: 8 p.

Thongsri, S. 1974. Economic evaluation of manual and mechanical land clearing and preparation: a case study of rubber replanting in Songkhla. Bangkok, Thammasat Univ. Thesis (M.Econ.):74p.

UNDP, 1978. Asian countries co-operating to upgrade natural rubber cultivation and processing. New York, UNDP/TCDC Case Study No. 23: 3 p.

Virgo, K.J. and Yaselmuiden, I.L.A., 1979. Cultivating soils of tropical steeplands. World Crops (Nov./Dec.): 316-321.

12
Forestry

"The forest is a peculiar organism of unlimited kindness
and benevolence that makes no demands for its subsistence
and extends generously the products of its life activity;
it affords protection to all beings, offering shade even
to the axman who destroys it."

--Gautama Buddha

Tropical timber is the world's second largest
internationally-traded commodity, valued at US$8 billion annually.
(Coffee, the largest, earned US$12.2 billion in 1981.) Since most
tropical timber is harvested from the wild, the need to balance the
multiple values of forests is increasing, including watershed
protection, soil conservation, habitats for rare and endangered
plant and animal species, homes for forest-dwelling tribal people,
climatic stabilization, industrial wood, fuelwood (firewood and
charcoal), and a wealth of other forest products.

Various predatory forms of tropical forest utilization
either extract a few species of exceptional value, damaging or
abandoning the remainder in the process, or--in the case of forests
with high densities of commercially valuable species (e.g.,
Dipterocarpaceae)--clearcut large areas and provide no
regeneration. On the other hand, many rational forest management
systems can effectively protect the environment and the natural
resource base, while providing significant economic benefits to
rural areas. Such systems can feature reforestation; watershed
management; social forestry; "shelterwood" and other
environmentally sound selective harvesting practices; agroforestry;
and village-based, mixed-species, multiple-use woodlots.

Reforestation and Afforestation: Reforestation and
afforestation (tree planting in cut-over forest and in hitherto
unforested areas, respectively) are generally environmentally sound
practices because they often redress formerly irrational land use.
To the extent that they provide products formerly extracted from
the forests, they alleviate pressures to destroy intact natural
forest, while protecting and generally improving the environment as
the trees mature. For these reasons, all forms of reforestation
and afforestation are strongly encouraged as environmentally
beneficial, whether their objectives include wood production,
socio-economic development, or soil and water conservation.

Of the many types of reforestation, single-species
plantations, while cheaper and more productive of wood than
heterogeneous plantations, are often more vulnerable to pests and
diseases. Monoculture plantations tend to exhaust certain soil

nutrients to a greater extent and create less habitat for wildlife than do mixed-species plantations. Monospecific eucalyptus plantations outside Australia, for example, support neither a plant understory nor many birds or other wildlife. These disadvantages can be greatly mitigated by planting several species together, or even by re-approximating the original forest composition. Indeed, the widespread practice of intercropping of legumes (e.g., Leucaena) and other synergists among the trees improves productivity markedly. New reforestation systems, involving more than one species and more than one product (as in agroforestry), need increased research and trials in order to become widely implementable.

Social Forestry: "Social" or "community" forestry is usually a small-scale endeavor, more directly beneficial to the local inhabitants than are industrial plantations (which often provide relatively low employment and serve markets remote from the local people) (World Bank, 1980). Social forestry is far more labor-intensive than most industrial plantations.

As a type of social forestry, village woodlots represent a vital and renewable source of fuel, lumber, and other forest products. Such forest plantations, when sited on critical watersheds, help to maintain reliable downstream water supplies, reduce flood damage, control soil erosion, and minimize sedimentation of dams and irrigation and navigation canals (as is being done in India's Kandi Watershed and Colombia's Upper Magdalena Pilot Watershed Management projects).

Social forestry can involve multi-purpose tree species that yield harvestable products. For example, gum arabic (Acacia senegal) production could be a national priority in several Sahelian countries as a drought relief measure. In Niger, gum production has been among the most successful drought relief measures. In Mauritania, gum production in 1972 was virtually the only employer of women in the country, being a successful source of income when grain and livestock production failed. Across semi-arid Africa, Acacia species have historically performed such a function in normal dry seasons. Acacia and many other drought resistant leguminous trees can also help prevent desertification, provide fuelwood and cattle feed, and enrich the soil with nitrogen.

Deforestation: Deforestation is here used to mean serious damage to or outright loss of natural forest and long-term conversion of once-forested areas to other forms of land use, such as cattle pastures (as in Amazonian Brazil and Colombia), agricultural colonization (as in Indonesia's transmigration program), or rubber, cocoa, and oil palm plantations (as in West Africa and Malaysia). Some commercial timbers occasionally may be extracted and sold as part of the conversion, but usually most of the wood is burned. Conversion of natural forest to tree plantations or other agricultural uses remains a serious environmnental concern.

In addition to outright land clearing, the commercial logging of tropical forests can also stimulate deforestation, resulting from destructive logging techniques, logging rates that greatly exceed rates of forest regrowth, and the subsequent slash-and-burn cultivation that often occurs when logging roads penetrate new areas and lead to "spontaneous colonization" by landless peasants.

Most tropical moist forests are characterized by a very high diversity of tree species, of which relatively few have high commercial value. These valuable species (with trade names like teak, mahogany, rosewood, and meranti) comprise the bulk of international trade. For example, in West Africa, about 10 tree species (out of more than 300) account for over 70 percent of total exports (Spears, 1979). Thus, commercial logging usually employs selective cutting techniques rather than clear-cutting (except for such products as woodchips).

If only the commercially desired species were affected, the result would be a modification of the original forest ecosystem, rather than its destruction or gross degradation. Unfortunately, modern selective logging techniques often damage much of the remaining tropical forest beyond recovery, far more than is the case in most temperate forests. One reason for this difference is that lianas and other climbing plants connect tropical forest trees with one other. Thus, when one tree is felled, it tends to break or pull down several others with it (NAS, 1980a). Moreover, the heavy machinery usually used inadvertently breaks limbs and scrapes bark off many tree species without commercial value; such injuries are often fatal to these trees, which are highly susceptible to attack by pathogens. In addition, this machinery often compacts or scrapes off the thin tropical forest soil, thereby preventing forest regeneration.

If more careful logging techniques were employed, the damage to the residual forest could easily be cut in half (UNEP, 1980). Examples of such techniques are to replace some of the heavy machinery with manual labor or elephants (still widely used in Burma), and to server the attached lianas prior to felling a tree. According to studies in Sabah, Malaysia, severing of lianas prior to selective cutting could reduce logging damage by 20 percent, at a very reasonable cost of US$2 per tree, (or 25 cents added to the price of each cubic meter of exported logs) (Hadi and Suparto, 1977). Indonesia has recently imposed a system of charging timber concessionaires US$4 for each non-marketed tree destroyed or damaged by careless logging practices (Soeriaatmadja, pers. comm., 1981). In most other countries, incentives are still lacking for logging companies to use any forest-conserving techniques, regardless of how slight the cost may be.

Prevailing differences of opinion concerning the value of undisturbed forest stem from Locke's fallacy, which maintains that forest has no value unless and until it is directly exploited by people. However, natural forest not only provides environmental services in perpetuity, but also safeguards a wealth of plant and animal species as a "genetic bank" for posterity. Many tropical forest-dwelling species have significant future economic potential as sources of medications, industrial products, and genetic inputs to agriculture, as well as aesthetic, spiritual, and cultural enrichment, if they are not first extinguished as a result of deforestation. For these reasons, tropical deforestation has been likened to burning the world's libraries for the sake of one winter's warmth (Chapter 24).

The greatest benefits of natural forests to humanity lie in the generally undervalued environmental services they provide, including flood control, groundwater recharge, protection of fluvial navigation, maintenance and restoration of soil fertility, control of landslides and erosion, prevention of sedimentation in waterways (including irrigation and other hydraulic systems), suppression of large fluctuations of plant and animal populations, buffering of the climate, promotion of hydrological cycling, and purification of air and water (by acting as a sink for carbon dioxide and various pollutants). Since these "free goods" are provided and maintained without charge to society by intact forests, their disruption will be expensive and damaging. Substitution of even tiny components of this environmental protection service--such as by flood and erosion control methods applied to deforested lands, and relief supplies for flood-ravaged communities--consumes inordinate quantities of resources and human energy that would be better applied elsewhere. This waste is entirely avoidable at low cost by leaving adequate protective forest intact.

The choice of the specific site for a proposed agricultural project is environmentally critical. Some sites should be scrupulously avoided altogether, such as those within territories occupied or used by vulnerable tribal or ethnic minorities (World Bank, 1982). Forests or other natural ecosystems should not be converted to agriculture if they occur within protected areas (such as national parks), or if they are known sites of rare or endangered species. Agricultural projects should be sited to the fullest extent possible on non-forest, cut-over, or degraded land, rather than in primary forest. Forestry plantation projects should be located on areas of already deforested land, rather than in areas of natural forest. If, however, environmental factors are outweighed by economic or other concerns, deforestation can be mitigated by the measures summarized in the following table (Figure 12.1).

Figure 12.1: Measures Mitigating Tropical Deforestation

1. Environmental reconnaissance aids in site selection. Protected areas, special habitats, wildlife refuge, or tribal homeland areas to be avoided are determined during project identification.

2. Ecosystem or biotic inventory provides an opportunity to collect (for museums or herbaria) or salvage (for relocation of live organisms) significant species and to study the tract before it is converted.

3. Methods of clearing are ranked by the degree of environmental effects in Figure 12.2.

4. Preservation of protective forest: Critical stands are preserved along watercourses, on slopes, and around reservoirs and irrigation headwaters. The size and location of such zones should be determined according to topographic features and the need for other uses, such as natural firebreaks.

5. More efficient use of cut forest: Broadening the spectrum of harvested species reduces the total area to be cut, while obtaining the same volume of wood. Promotion of less commercially-used species to the fullest extent possible assists greatly. Through improved technology, it is possible to recover parts of trees that previously were discarded (e.g., branches, stumps, and sawmill wastage). Conservation-oriented technologies, such as more efficient wood-burning stoves and simple timber preservation techniques at the village level, also help to conserve the remaining forest resources.

6. Reforestation, if effectively executed and significant in scope (near the project or elsewhere) can greatly mitigate the effects of deforestation, even in the same project. Since plantation forestry can be up to ten times more productive biologically than logging of natural forests, promotion of tree plantations can reduce deforestation pressures.

7. Compensatory preserves: The setting aside of protected forest areas comparable to the deforested tract helps to compensate for the loss. Compensatory project components may include strengthening or augmenting the national park system. Such compensatory wildland management areas may be sited near the project area, adjacent to existing preserves, or elsewhere, as long as they protect an ecosystem similar or identical to the one being converted (World Bank, 1983).

8. Research on sustained-yield agro-ecosystems can be included as a project research component to promote harvest systems from undisturbed forest that would leave the vegetation intact or as little disturbed as possible, thereby minimizing the loss of natural forest habitat. Education, training, and extension of the results of this research will be essential if sustained-yield agricultural systems are to be achieved within the carrying capacity limits of tropical forest regions (Janzen, 1973).

9. Recycling of wood, wood products, and waste paper, while less immediately related to deforestation than the above measures, could, if undertaken energetically, moderate the demand for "virgin" pulp and thus alleviate pressures on forests.

The World Bank's Forestry Sector Policy Paper (1978) clearly specifies forest protection as a high priority, emphasizes the environmental benefits of forestry projects, and warns of the dangers of neglectful deforestation. Where a project designer finds some deforestation unavoidable, including such measures as those outlined above will minimize the resulting environmental damage.

Land Use Criteria: As precautions to avoid environmental change are necessarily site-specific, only general options can be suggested here. This book focuses on the tropics--areas with warm climates and rapid soil chemical processes. The higher and the more intense the rainfall, the greater the risk of erosion and leaching of critically limiting mineral nutrients and loss of topsoil. Only woody perennial tree crops and forests confer adequate soil protection when annual precipitation exceeds 2.5 meters. To clear intact forest in areas receiving 3 meters or more of rain invites damage to most poor tropical soils. Where there is a regular dry season, the potential for damage is less, especially to the extent that crops can become established before the onset of the next wet season.

Since soils vary greatly (both physically and structurally), in their inherent susceptibility to erosion and leaching (irrespective of vegetative cover), appropriate land use can be determined only after soil reconnaissance. Once the rainfall, length of dry season, slope, and soil type are known, appropriate specific criteria can readily be outlined.

Precise specification of acceptable land use within given slope limitations can be effective, provided there is adequate field monitoring. For example, both in Malaysia and the Philippines, no cultivation is allowed (although some occurs) on gradients above 18^0 (or 1:3 slopes), while terracing is essential on 1:5 to 1:3 slopes. Gentle slopes up to 5^0 (1:10) are not as susceptible to accelerated erosion (except occasional gullying), except in areas of very heavy rainfall. Susceptibility to erosion increases almost geometrically on slopes between 5^0 and 18^0. Some agencies permit clear-cutting on slopes up to 12^0, but only selective logging above 12^0. Indonesian transmigration regulations mandate perennial or tree crops for all land of 8^0 slope and over. Row crops on slopes of 3 percent or steeper are susceptible to high erosion rates.

Methods of Deforestation: The methods by which trees are removed from a tropical forest tract greatly influence the future productivity of the site. Most nutrients are in the vegetation rather than in the soils. Sustained yields can be obtained by carefully plucking out individual trees of selected species and allowing adequately long periods of regeneration. The shelterwood system, forest enrichment or refining, liberation thinning of natural vegetation, and enrichment planting to improve regeneration are also environmentally protective, long-term management systems.

Where the project involves conversion of forest to a
different form of land use, such as a rubber plantation, the gradual
enrichment planting of the rubber (as has been practiced
successfully--Dawkins, 1955), and the cautious utilization of other
trees will reduce damage. If this is judged to be difficult or
unduly slow, then forest clearance options ranking high on the list
shown in Figure 12.2 will result in less damage than those at the
bottom of the list. The main variables are the extent to which the
topsoil is scraped off (e.g., by the bulldozer blade or by a winched
tree scoring the ground) and the soil compacted and churned by heavy
machinery. These relate to the weight, type, and turning radius of
the machinery, the skill of the operator, and the nature of the soil
itself. A cost-effective choice must be made between (a) heavy
machinery powerful enough to cut trees without several churning
passes, (b) lighter, less powerful machinery which may have to try
several times to cut the tree, or (c) people with portable or
hand-held winches, chainsaws, or axes (cf. Ross, 1980). The Jari
Gmelina and Pinus pulp project in Amazonia reversed mechanized
clearing policies when subsequent wood yields were very low due to
soil compaction and removal of the thin organic layer of topsoil.
In major projects involving deforestation (e.g., Indonesia's
Transmigration Program and Ghana's oil palm project), bulldozers are
being abandoned in favor of manual clearing to protect the soil.

Subsequent agricultural productivity varies greatly,
depending in part on the method by which land is cleared. Sanchez,
et al. (1977) compared the effect of traditional slash-and-burn
clearing with bulldozer land clearing in one area of the Peruvian
Amazon:

"The favorable slashing and burning method produced more
favorable changes in soil properties and crop yields than
bulldozing. Burning increased the supply of exchangeable
bases and available soil phosphorus severalfold, decreased
aluminum saturation, and retarded the organic matter
decomposition process by about six months. In sharp
contrast, the bulldozed areas suffered from severe soil
compaction, did not receive additional bases, remained high
in aluminum saturation levels and had available phospate
and potassium values below the critical levels. Yields of
upland rice, cassava, maize, soya beans, and guinea grass
were consistently superior in the burned clearings.
Considering that bulldozer clearing costs are about four
times more than slashing and burning, both agronomic and
economic parameters suggest the use of the traditional
method in the transition from shifting to continuous
cultivation."

Figure 12.2: Environmental Ranking of
Tree Felling and Removal Methods
(F = Felling; R = Removal)

Minimum Environmental Damage	Method
Hand removal	R
Blimp, balloon, or helicopter	R
Aerial cable	R
Manual felling	F
Chainsaw	F
Elephant	R
Other draft animal	R
Hand winch	R
Mechanical winch	R
Shearing blade	F and R
Tree pusher	F and R
Skidder	R
Tractor(s) and chain	F and R
Bulldozer (various types)	F and R

Maximum Environmental Damage

Note: These methods are ranked in approximate order of
environmental preference; it is clear that economics, labor
supply, terrain, and other criteria will also influence the
final choice. The environmental effects will differ
according to whether trees are extracted from even-aged
plantations or uneven-aged natural forests.

The various means for disposing of unusable forest remnants
also have differing environmental consequences. On-site decay
simultaneously retains nutrients and protects the ground, but this
may be too slow or obstructive for the project. On-site burning
recycles nutrients (except nitrogen and sulfur) fairly evenly, but
often this is not feasible because the slash is too wet to burn.
Piling waste into windrows and allowing it to dry for eventual
burning is a common method, but the ash then remains unevenly
distributed. Chipping and allowing the chips to decay avoids the
volatilized loss of nitrogen and sulfur inherent in any burning.

The methods by which windrowing is achieved are subject to
environmental considerations similar to those encountered during the
original clearing. Contouring of the windrows reduces rain damage
and surface erosion. The distance between windrows and their height
can be adjusted to the rainfall and slope. Windrowing into gullies
disrupts drainage and promotes loss of nutrients.

Finally, the timing of the clearing operation and of
initial treatment after clearing are important. Cleared areas left
unprotected in a wet season will lose more nutrients than during a
dry season. Cleared tracts that are immediately planted with a
cover crop (including legumes and vines) lose fewer nutrients and
organic matter than do unprotected tracts.

Habitat for Forest-Dwelling Species: In addition to its environmental protection services, intact forest also serves as a wildlife habitat or gene bank (Chapter 24). Some environmental protection services can be approximated by tree plantations, albeit expensively and not as effectively. Most habitat or gene bank values cannot be approximated as they exist only in natural forest and not at all in tree plantations, even those of mixed species. Establishing compensatory preserves and strengthening any existing system of national parks and similar protected areas are the only mitigatory measures for the loss of natural habitat that accompanies deforestation.

Although the habitat or gene bank value of forest is even more difficult to quantify economically than the environmental service value, it is nonetheless very real. The United States' "Global 2000" study ranks species extinction as an urgent problem that is "without precedent in human history" (CEQ, 1980). Tropical moist forests are biologically the richest ecosystems on earth, containing 40 to 50 percent of the world's species on 6 percent of its land area (Ledec, 1983). Many of these species have very localized distributions, and are thus vulnerable to sudden extinction as a result of habitat loss (more so than most temperate zone species) (UNEP, 1980b). A significant number of species, and even entire genera, can be extinguished by one poorly-planned, large-scale deforestation project.

Although impossible to quantify accurately, the social costs of species extinctions also are very real. Aside from the ethical and aesthetic losses inherent in the voluntary extermination of unique life forms, the loss of many currently unknown species can severely undermine humanity's future ability to develop new medications, industrial products, agricultural systems, and other items of great economic value (Myers, 1983; Prescott-Allen, 1982). Consequently, projects should be planned in a manner that avoids causing the extinction of any species. Extinctions can be avoided or minimized by careful project siting and design, and through compensatory components such as augmenting a national park system (Figure 12.1) (see World Bank, 1983).

Plot Size: Conversion of tropical moist forest for agricultural settlement or colonization is environmentally undesirable (Figure 12.3), since the forest is irreversibly lost and the agricultural production is usually ephemeral; about three harvests are possible before weeds and declining yields make it preferable to cut more forest. Since this realization has yet to be accommodated in many land settlement projects, the more people settled per hectare, the less the damage and the more efficient the use of land resources will be. In Amazonia, 100-hectare plots are common in settlement schemes. The only way one family can manage such a large parcel of real estate without mechanization (which it usually cannot afford) is to burn down the forest and run a small herd of cattle, which is especially undesirable and unsustainable. By contrast, in the Outer Islands of Indonesia, a family with 2-3 hectares of very similar land often meticulously manicures it by terracing any creeks for rice, constructing fish ponds, and planting home gardens (**pekarangan**--Chapter 11) and orchards (Goodland, 1981, 1983).

Figure 12.3: **Environmental Ranking of Tropical Wet Forest Utilization Options**
Note: This is **not** an economic ranking.

(The more desirable and least risky options head the ranking, followed by the less desirable and riskier options.)

1. **INTACT FOREST**

   1.1 Biological reserve; scientific repository; gene pool, germ plasm, habitat, phytochemical, and ethnobotanical resources.

   1.2 Environmental protection services: climatic buffer, watershed protection, and protection of downstream activities.

   1.3 Reservations for indigenous peoples based on natural, legal and moral criteria; also for knowledge of indigenes.

   1.4 National park development; national and international tourism; non-consumptive forms of recreation.

   1.5 Collecting, gathering, and tapping of forest plants; game and fish culling.

2. **UTILIZATION OF NATURAL FOREST**

   2.1 Sustained-yield management (as in the Nigerian and Malaysian shelterwood forestry systems now in disuse).

   2.2 Obtaining leaf protein and leaf and other chemicals without killing the tree.

   2.3 Selective felling with careful removal.

   2.4 Bole removal, but with slash, roots, stump, bark, and branches left on the site, rather than removal of the whole tree.

   2.5 Enrichment planting, refining, liberation thinning, reconstitution management, and directed (artificial) regeneration.

   2.6 Clear-cutting small tracts, leaving adequate natural forest nearby for regeneration, or actually replanting.

3. TREE PLANTATIONS

3.1 Mixed-species polyculture products (e.g., rubber, oils, nuts, resins), rather than monoculture products.

3.2 Mixed-species polyculture timber plus synergistic species and products.

3.3 Monoculture timber, including lumber, veneer, plywood, particleboard, woodchips, pulp, fuelwood, and hogfuel.

4. AGRO-FORESTRY

4.1 Multiple-dimension forestry, e.g. "three-dimensional" forestry with timber, tree products, synergists, browse, production understory components, and graze.

4.2 Polycropping and intercropping of, e.g., rubber and synergists with understory and annuals.

4.3 "Taungya system": annuals and perennials planted simultaneously, eventually becoming tree plantations; chena.

4.4 Treed pasture: wood and other tree products plus synergists; browse and multi-species graze (e.g., legumes, forbs, and grasses).

4.5 Subsistence rotation gardens, e.g., Indonesian pekarangan, Mayan "home gardens", Sri Lankan "Kandy gardens", and Mexican chinampa systems involving trees, perennials, and annuals, along with small livestock and fishponds.

5. AGRICULTURE

5.1 Long fallows; small cultivated areas; multiple varieties of crop species; selective breeding for tolerance of pests and infertile soils; crop rotation.

5.2 Perennial crops are preferred over annuals; some land settlement schemes adopt this model.

5.3 Subsistence crops are preferred over export and cash crops such as tobacco or sugar.

5.4 Oligotrophic export crops, containing mostly hydrocarbons or carbohydrates, are preferred over eutrophic exports.

5.5 Multi-species pasture for mixed herbivores; small livestock and stabled cattle.

5.6 Oligoculture pasture for monospecific herbivores, especially extensive ranching for cattle export, is the least environmentally desirable option under prevailing, low-intensity management practices.

Future Uses of Converted Forest: The notorious record of deforestation through cattle ranching on infertile soils in high temperature, high rainfall areas of low technology and management has now been carefully documented by Hecht (1981). In general, the many and repeated attempts at large-scale, continuous cash crop (annuals) and export cattle production in the formerly forested tropics have failed (with certain exceptions). One recent colloquium prioritizing constraints to Amazonian development dwelt almost exclusively on difficulties of credit and land tenure, poor administration, deficient infrastructure, weak governmental support, inadequate market demand, and so on. The general failure of the colloqium's participants to acknowledge the region's overriding environmental constraints (such as infertile soils, pest and weed pressure, ecological interdependence) is striking and unfortunate (cf. Jordan, 1982). High-input agriculture (involving petroleum, chemical fertilizers, and biocides) is being questioned even in rich countries based in temperate zones with relatively fertile soils and less pest pressure; it is likely to be even less appropriate for the humid tropics.

Although the infertile nature of most wet tropical upland soils has long been realized by soil scientists, how the luxuriant jungle manages to thrive is only now being elucidated. Jordan and his colleagues are analyzing the above-ground closed nutrient cycle of the tropical moist forest, which is so leak-proof that incoming nutrient rates (mostly through rain and dust) closely balance outgoing nutrient rates (through leaching and streamflow). Even forest canopy leaves scavenge nutrients from the rain, and roots grow out of the soil up tree trunks to intercept nutrients flowing down the bark. The implication of this ecosystem adaptation to infertile soils is that the forest cannot be clear-cut and burned over large areas, and then be expected to produce crops for more than a few years.

Sanchez, et al., (1982) found that forest can be replaced by annual cropping systems using imported chemical fertilizer for eight years in Amazonian Peru. However, this approach neglects the vast cost of building nearly 2,000 km of highway across the Andes and its onerous maintenance costs, the fact that fertilizers are heavily (and inefficiently) subsidized by the Peruvian Government, and the need for costly biocides. Most Amazonian farmers who allowed Sanchez' high input experiment to be conducted on small parts of their property have not adopted these methods. The inapppriateness of reliance on expensive inputs (e.g., fertilizers, biocides, tractors, and diesel) may now have been perceived, since Sanchez' program has recently begun research into low-input farming techniques.

Scientists and farmers, but unfortunately not yet most economists and politicians, are realizing that whatever sustained-yield may be obtainable from most tropical moist forest soils, the yield will probably be so small that it is preferable to leave the forest intact and support growing populations with food produced more efficiently on existing agricultural lands. Development alternatives for tropical moist forest that are relatively sustainable and environmentally benign (e.g., pekarangan home gardens, fish ponds, orchards, and the chinampa system) should be strongly promoted in place of current practices, but even these superior methods are unlikely to support dense human populations in humid tropical areas with characteristically poor soils (Janzen, 1973; Goodland, et al., 1978; Fearnside, 1983 and references in Chapter 11).

Fuelwood: Wood is the only source of domestic energy for millions of the very poor. In the aggregate, wood fuels in developing countries probably account for about 85 percent of all non-commercial energy, other than human and animal energy (NAS, 1980). As wood can be a renewable resource, projects designed to provide fuelwood (firewood and charcoal) on a sustained-yield basis are environmentally desirable. "By the turn of the century, at least a further 250 million people will be without wood fuel for their minimum cooking and heating needs and will be forced to burn dried animal dung and agricultural crop residues, thereby further decreasing food crop yields" (Spears, 1978). Project design should assess the tradeoffs between growing wood for fuel, using animal dung for fuel, or using the dung for fertilizer or biogas generation and the residuals for fertilizing food crops.

The leguminous tree crop **Leucaena leucocephala** ("Ipil-Ipil") can produce in excess of 50 cubic meters of wood/ha/year and can grow on steep slopes and marginal soils (NAS, 1980b). Several hundred promising fuelwood species are listed in the NAS (1980b) "Firewood Crops" report, which details the advantages of 60 of them. The more species-diverse the plantations, the fewer the environmental risks.

Other underutilized trees can greatly benefit different areas. FAO has called the Neem Tree (**Azadirachta** indica, of the Meliaceae family) "the greatest boon of the century". Neem thrives on and enriches impoverished soil. The leaves (pH 8.2) help to neutralize acid soil, while adding nitrogen. The tree contains several non-toxic insect repellents in such quantities that Neem protects other crops grown nearby. Virtually every part of Neem is useful, including the seeds, bark, wood, and oil. However, the species remains relatively unknown outside India and West Africa.

Mangroves: Mangroves (saltwater swamp forests) are environmentally and economically important, not only for fuel, resin, tanning bark, termite-resistant lumber, and other products, but also for stabilizing shorelines and preventing saltwater intrusion into freshwater supplies. Possibly their most significant service is the provision of shelter and food for shrimp, shellfish, and fish (even species caught far offshore), many of which almost totally depend upon mangroves during some stage of their life cycle. Even the most economically important fresh water shrimp

(Macrobrachium) depends upon mangroves in its larval stage (Linden and Jernelov, 1980; Rollet, 1981; Saenger, et al., 1983). Mangroves act as a chemical and physical filter, retaining nutrients that are washed from the land and deposited as silt. The nutrients are flocculated into particles with mud, preventing their rapid loss. These nutrient-rich particles are released slowly but steadily into estuarine waters, providing a constant, reliable supply of food for marine organisms. Upstream cutoffs of nutrients, as by dams for irrigated rice, often damage mangroves. Landfills between mangrove stands and the upland areas can kill extensive tracts of these trees (Chapter 16).

References

AID, 1983. Common fuelwood crops (tropical). Washington, D.C., AID/USDA Forest Service, 354 p.

Agroforestry, 1979. Symposium on tropical agriculture. Amsterdam: Royal Tropical Institute, Department of Agricultural Research, 47 p.

Ashton, P.S. 1981. Techniques for the identification and conservation of threatened species in tropical forests (155-164) in Synge, H. (ed.) Biological Aspects of Rare Plant Conservation. London, Wiley.

Assessing tropical forest lands: their suitability for sustainable uses. 1980. Honolulu, University of Hawaii, East-West Center, 69 p.

Auchter, R.J. (ed.). 1978. Proceedings of a conference on improved utilization of tropical forests. Madison, Wisconsin, USDA Forest Service, 569 p.

Beall, H.W., Bene, J.G., and Cote, A. 1977. Trees, food, and people: land management in the tropics. Ottawa, IDRC, 52 p.

Caterpillar Tractor Co. 1974. The clearing of land for development. Peoria, Illinois, 111 p.

Caterpillar Tractor Co. 1979. Caterpillar performance handbook. Peoria, Illinois, 480 p.

Caufield, C. 1982. Tropical moist forests: the resources, the people, the threat. London, Earthscan, 68 p.

CEQ, 1980. The Global 2000 Report to the President. Washington D.C., U.S. Council on Environmental Quality, I: 47 p.

Chijicke, E.O. 1980. Impact on soils of fast-growing species in lowland humid tropics. Rome: FAO Forestry Paper 21: 111 p.

Christensen, B. 1983. Mangroves--what are they worth? Unasylva 35(139):2-15.

CITES, 1974. Convention on international trade in endangered species of wild fauna and flora. Gland, Switzerland: CITES secretariat (v.p.).

Combe, J.H., Jimenez-Saa, H., and Monge, C. 1981. Bibliography on tropical agroforestry. Turrialba, Costa Rica: Centro Agronomico Tropical de Investigacion y Ensenanza (CATIE), 6:67 p.

Dawkins, H.C. 1955. The refining of mixed forest. Empire Forestry Review 34:199-191.

Eckholm, E. 1979. Planting for the future: forestry for human needs. Washington, D.C., Worldwatch Paper 26:64 p.

FAO, 1975a. Manual of forest inventory with special reference to mixed tropical forests. Rome, FAO:200 p.

FAO, 1975b. The methodology of conservation of forest genetic resources: a report on a pilot study. Rome, FAO:127 p.

FAO, 1978a. Forestry for local community development. Rome, FAO Forestry Paper 7:114 p.

FAO, 1978b. China: forestry support for agriculture. Rome, FAO Forestry Paper 12:103 p.

FAO, 1982. Environmental impact of forestry. Rome, FAO Conservation Guide 7:85 p.

Fearnside, P.M. 1983. Development alternatives in the Brazilian Amazon. Interciencia 8(2):65-78.

Fearnside, P.M. 1979. The development of the Amazon rain-forest: priority problems for the formulation of guidelines. Interciencia 4(6):338-343.

Fox, G.D. 1973. Technological opportunities for tropical forestry development. Washington, D.C., USAID, DSDS/OST:31 p.

Goodland, R., Irwin, H.S., and Tillman, R. E. 1978. Ecological development for Amazonia. Sao Paulo, Ciencia e Cultura 30(3):275-289.

Goodland, R. 1981. Indonesia's environmental progress in economic development (215-276) in Sutlive, V. H. et al. (eds.) Deforestation in the Third World. Williamsburg Va., College of William and Mary, Studies in Third World Societies, 278 p.

Goodland, R. 1983. Environmental progress in Amazonian development. in Hemming, J. (ed.) Change in the Amazon Basin. Manchester, Manchester University Press:2 vols.

Hadi, S. and Suparto, R.S. (eds) 1977. Long term effects of logging in South-East Asia. Bogor, Regional Center for Tropical Biology.

Hecht, S. B. 1981. Cattle ranching in the Amazon: analysis of a development strategy. Berkeley, Univ. California Geography Dept. (Dissertation): 450 p.

ICRAF, 1982. A selected bibliography of agroforestry. Nairobi, International Council for Research in Agroforestry, 60 p.

IUCN, 1975. The use of ecological guidelines for development in tropical forest areas of Southeast Asia. Morges, Switzerland, IUCN Publications, new series No. 32:185 p.

IUCN, 1975. The use of ecological guidelines for development in the American humid tropics. Morges, Switzerland IUCN Publications, new series No.31:249 p.

Janzen, D. H. 1973. Tropical agroecosystems. Science 182: 1212-1219.

Jordan, C. F. 1982. Amazon rain forests. American Scientist 70:394-401.

Lal, R. 1981. Clearing a tropical forest II: effects on crop performance. Field Crops Research 4(4):345-354.

Leach, G. 1976. Energy and food production. Guildford, IPC publ. 137 p.

Ledec, G. 1983. The political economy of tropical deforestation in Leonard, H. J. (ed.) The politics of environment and development. New York, Holmes and Meier (in press).

Linden, O. and Jernelov, A. 1980. The mangrove swamp - an ecosystem in danger. Ambio 9(2):81-88.

Maason, J.L. et al. 1979. Demonstration of integrated management and utilization of tropical forests. Plan for managing the Alexander von Humboldt National Forest, Peru. Rome, FAO/UNDP:3 vols.

Myers, N. 1983. A wealth of wild species. Boulder, Co., Westview Press, 274 p.

NAS, 1980a. Conversion of tropical moist forests. Washington, D.C., National Academy of Sciences, 205 p.

NAS, 1980b. Firewood crops: Shrub and tree species for energy production. Washington, D.C., National Academy Sciences, 237 p.

NAS, 1982. Ecological aspects of development in the humid tropics. Washington, D.C., National Academy Sciences, 297 p.

NRC, 1980. Research priorities in tropical biology. Washington, D.C., National Research Council, 116 p.

OTA, 1983. Technologies to sustain tropical forest resources. Washington, D.C., Office of Technology Assessment (US Congress): 2 vols.

Pirie, N.W. 1969. Food resources: conventional and novel. Harmondsworth (UK), Penguin, 208 p.

Poore, M.E.D. 1976. Ecological guidelines for development in tropical rain forests. Gland, IUCN:39 p.

Poulsen, G. 1978. Man and trees in tropical Africa: three essays on the role of trees in the African environment. Ottawa, IDRC:31p.

Prescott-Allen, R.& C. 1982. Economic contributions of wild plants and animals to developing countries. Washington, D.C., USAID/MAB: 96 p.

Rollet, B. 1981. Bibliography on mangrove research 1600-1975. Paris, UNESCO:479 p

Ross, M. R. 1980. The role of land clearing in Indonesia's transmigration program. Bull. Indonesian Econ. Stud. 16(1):75-87.

Saenger, P. et al. (eds.) 1983. Global status of mangrove ecosystems. The Environmentalist (Suppl) 3: 88 p.

Sanchez, P.A., Bandy, D.E., Villachica, J. H., and Nicholaides, J. J. 1982. Amazon basin soils: management for continuous crop production. Science 216:821-827.

Sanchez, P.A., Seubert, C.E., and Valverde, C. 1977. Effects of land clearing methods on soil properties of an ultisol and crop performance in the Amazon jungle of Peru. Tropical Agriculture 54 (4): 307-321.

Schubart, H. and Salati, E. 1980. National resources for land use in the Amazon region: the natural systems. CIAT, Cali, Colombia:50 p.

Smith, N.J.H. 1981. Fuel forests: a spreading resource in developing countries. Interciencia 6:336-343.

Spears, J.S. 1978. Wood as an energy source: the situation in the developing world. Amer. Forestry Assoc.

Spears, J.S. 1979. Can the wet tropical forest survive? Commonwealth Forestry Review 58(3):165-180.

Spears, J.S. 1983. Saving the tropical forest ecosystem: a discussion paper. Washington, D.C., The World Bank (Harvard Univ. Congr. Land Policy) (ms).

UNEP, 1980a. Mountain ecosystems:a review. Nairobi, UNEP Report No. 2:38 p.

UNEP, 1980b. Tropical woodlands and forest ecosystems:a review. Nairobi, UNEP Report No. 1:84 p.

UNESCO/MAB, 1972. Ecological effects of increasing human activities on tropical and sub-tropical forest ecosystems. Paris, UNESCO/MAB Report 3: 35 p.

UNESCO/MAB, 1974. Ecological effects of increasing human activities on tropical and subtropical forest ecosystems. Rio de Janeiro, Report 16:96 p.

UNESCO, UNEP, FAO. 1978. Tropical forest ecosystems: A state of knowledge report. Paris, UNESCO, 683 p.

World Bank, 1978. Forestry Sector Policy Paper. Washington, D.C., World Bank, 63 p.

World Bank, 1980. Sociological aspects of forestry project design. Washington, D.C., World Bank AGR Tech. Note 3:100 p.

World Bank, 1982. Tribal peoples and economic development. Washington, D.C., World Bank:111 p.

World Bank, 1983. Wildlands management in economic development: a policy proposal. Washington, D.C., The World Bank, Office of Environmental Affairs (draft ms).

Wyatt-Smith, J. 1979. Agro-forestry in the tropics: a new emphasis in rural development. Span 22:65-27.

13
Large Livestock:
Cattle and Buffalo

"... a gross misuse of resources. Livestock are turned into 'reverse protein factories' by feeding them protein-rich grain, legumes, and fish-meal in order to produce a rather smaller amount of protein in the form of meat." --Stanley, IDRC, Ottawa, 1983.

"The quality of a culture is measured by its reverence for all life."--Albert Schweitzer.

Overview of Animal Protein: Vertebrate protein can be ranked, in very general terms, from the environmentally most desirable to the least desirable--that is, from fish (both marine and freshwater), through small livestock, to large livestock (which tend to be the most problematic). Invertebrate protein, which includes insects, molluscs, crustaceans, and micro-organisms, is not further detailed in this book. In most instances, plant, invertebrate, and aquatic forms of protein are strongly preferred environmentally over mammalian flesh for the reasons outlined below.

Many, if not most, tropical people are induced by their environment, by poverty, or by cultural and religious preferences towards a great reliance upon plant protein and even vegetarianism. Meat, particularly from intensively-produced mammals, is too expensive for much of the world's population. It is expensive largely because its production makes inefficient use of energy, nutrients, and other resources. Since relatively few people (i.e., the affluent) profit from intensive livestock production, it is less socially equitable than plant protein production (Figure 13.1). The world food situation and the global renewable resource base would be enhanced to the extent that wastage is reduced in agricultural practices. For this reason, it is desirable to shorten the food chain and increase natural resource efficiency by promoting vegetable over animal production, and fish and fowl over mammalian protein (with particular reservations against grain-fed beef). To the extent that scarce natural resources are inefficiently used and world hunger remains unalleviated, production of such luxuries as grain-fed beef is difficult to justify socially.

A society's or individual's position in the existing spectrum between largely carnivores (e.g., the Inuit nation), omnivores, vegetarians, and vegans (e.g., the Jains of India) is determined by food availability, affordability, religion, ethics, custom, and other factors. In addition, people concerned with species extinctions may avoid beef raised in tropical forest pastures. People concerned with natural resource use efficiency and feedlot-related pollution may avoid grain-fed beef, as well as moving down the food chain to increased vegetarianism. Those concerned with their personal health may reduce their consumption of marbled beef (due to its high saturated fat and cholesterol content). Those concerned with cruel treatment of animals may eschew veal or strasbourg goose liver pate. In such a complex arena of conflicting value judgements, there is no single answer. Opinions differ, for example, on raising permanently caged, semi-tame animals only for ultimate slaughter, versus the killing of wild animals. How humane is "humane" slaughter, or the cage rearing of hogs or commercial chickens under cramped conditions? Until we are better able to influence the sex ratios in livestock, it is impossible to promote eggs and milk without considering the disposition of male animals.

However, the world's ethical perspectives continue to change and to clarify such issues. For example, in most countries, flogging one's horse to death is no longer socially sanctioned. Mass killing of game animals is no longer sanctioned in tsetse control for cattle projects (Chapter 19). Some World War II prisoners of war in camps felt able to eat cats; others did not. It is worthwhile to respect the many societies in which it is unethical to treat animals as if they have no intrinsic value. To treat animals as if they have instrumental value only for human ends can be construed as arrogant and perhaps immoral (Regan, 1982; Singer, 1975).

Beef Cattle as Protein: The ruminant's extraordinary virtue of harvesting and then converting cellulose to quality protein can be efficiently exploited wherever native grasslands remain unsuited to tillage. The environmental value of livestock in general is their recycling of agricultural waste (such as from milling industries and crop residues), thus alleviating the chronic problem of disposing of farm surpluses. Their dung is highly valued as building material, manure, fertilizer, and fuel in many countries.

The use of livestock to recycle small surpluses of grain beyond that which the individual farmer can sell or store can improve efficiency. However, beef cattle are creating major environmental problems because their production recently has become such a central feature of Western agriculture. Only since the 1950's has beef consumption exceeded pork and mutton. Dietary efficiency, the proportion of animal products in the diet, is related to income levels. The steer is the least efficient of all livestock at converting grain protein to animal protein: an average steer requires up to 8 kg of grain to produce 1 kg of meat (Figure 13.3). A larger portion of the energy and nutrients is lost in manure (unless recycled) or in the carcass, and much of the energy is metabolized by the animal during conversion. Approximately half of the world's primary protein commodities are fed to livestock, which are among the least efficient heterotrophic converters (Figure 13.3). Reduction of intensive livestock production would reduce

Figure 13.1: The Fate of the World's Protein Resources

Commodity	Average Percent Protein Content of Dry Matter	Percent Fed to Livestock
Grain	8-14	33-35
Oilseeds	26-40	60-75
Fish	15-25	40-50
Milk Products	3-33	25-40

Source: USDA Handbook--Consumption and Utilization of Agriculture Products.

Figure 13.2: Environmental Ranking of Cattle Production Methods

Ranking	Method	Grain Content	Remarks
Most Desirable	1. Stall-fed on cut grass and fodder.	No grain.	Least damaging to land.
	2. Natural range.	No grain.	Sustainable if not overgrazed.
	3. Finishing on feedlots or improved pastures (sometimes fertilized and irrigated).	Partly grain.	Partly wasteful of grain
	4. Feedlots (U.S.-type).	Mainly grain.	Least efficient conversion of grain.
Least Desirable	5. Induced pasture.	No grain.	Deforestation; usually unsustainable.

Figure 13.3: Conversion Ratios of Grain into Animal Products

Animal (product)	Feedlot Conversion Ratio 1/
Beef (Steers)2/	8:1
Pork	6:1
Turkey	4:1
Eggs	3:1
Chicken	3:1
(Shellfish)3/	(ca. 1:1)

Source: USDA Economic Research Service.

1/ (Kilograms of grain and soy fed to produce 1 kg of meat, poultry, or eggs.

2/ Other livestock are more efficient converters than steers. Simpson and Farris (1982) note that the 8:1 ratio drops to 3:1 if one subtracts the dam and sire since they probably are not grain-fed, and calculate the ratio on the 14 to 16 month life, not just on the last grain-fed 4 to 6 months.

3/ In their natural habitat; for comparison only.

- 108 -

Figure 13.4

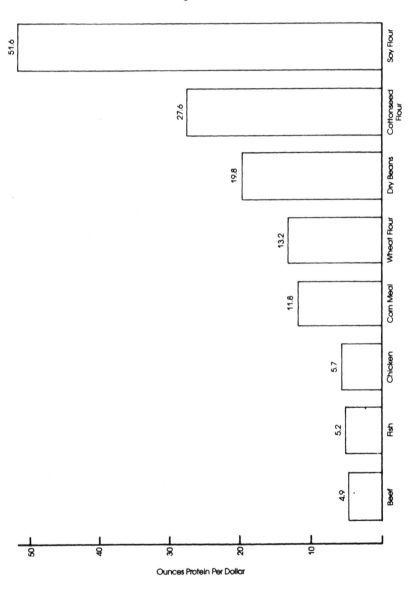

Figure 13.5

WORLD CEREAL CONSUMPTION,
AS FOOD AND FEED

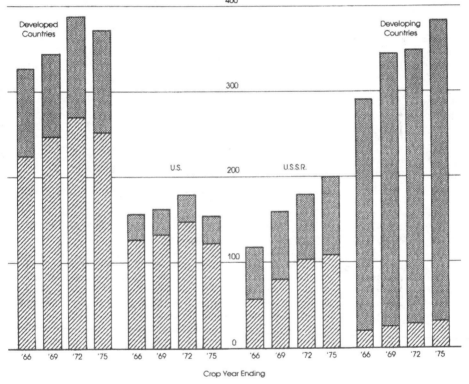

Food

Feed

Crop Year Ending

World Bank—25036

grain prices, hence may increase the availability of grain to the poor. Price ratios between feed concentrates and beef products are unlikely to make intensive beef production economically feasible in the tropics (Schaefer-Kehnert, 1981). In most tropical countries, beef is likely to be produced more cheaply under less intensive production systems which are based on grazing and forage crops (Figure 13.2).

By comparison, an acre of cereals can produce twice to ten times as much protein as an acre devoted to beef production; legumes can produce ten to twenty times more; and leafy vegetables twice to twenty times more. When fed to cattle, one acre of grain yields 77 days of person's protein requirements. By contrast, the same acre producing whole wheat flour provides up to 5,463 days of a persons's protein requirements, and one acre of edible soybean, would provide 15,100 such days. The UN World Food Council (Mexico, 13 June 1978) summarized the situation by pointing out that, "Ten to fifteen percent of the cereals now fed to livestock in developed countries is enough to raise the world food supply to adequate levels." If the resources invested in luxury cereal production for livestock feed were applied instead to tropical agriculture, they would vastly alleviate global hunger.

In developed countries, ten times as much fossil fuel energy is used to produce a unit of animal protein as to produce a unit of plant protein. One of the highest energy costs is for beef, produced under feedlot and range conditions requiring inputs of up to 78 kcal of fossil energy per kcal of usable beef protein. The disadvantages of such livestock production are compelling from the global energy budget alone (Chapter 21). By comparison, 36 kcal of fossil fuel energy are needed for each kcal of milk protein; for egg protein the ratio is 13:1.

According to the United States National Academy of Sciences (1982), Hegsted (1979) and others, meat is unnecessary for a healthy diet. Heavy consumption of beef, especially the grain-fed, higher grade marbled prime, may actually promote heart disease, cancer, and other ailments. Furthermore, meat production (involving feed production, watering, animal care, and meat preparation) requires 20 times as much water as plant protein production alone. The 100 kg of excrement produced per kilo of beef in feedlots imposes a heavily polluting burden on waterways unless it is recycled as fertilizer. Erosion is vastly greater in pastures than on forest in the tropics. Table 13.6 ranks livestock in general terms of their benefit to the environment and reflects the above and similar concerns.

Table 13.6: Livestock: Environmental Preference
(Approximate and Partial Ranking)

Rank	Animal	Food	Product	Preference
1.	Poultry	Residues, wastes	Meat, eggs	Recommended
2.	Swine	Residues, wastes	Meat	Recommended
3.	Sheep	Natural range 1/	Wool, meat	Recommended
4.	Wildlife or goats	Natural range 1/	Meat, hides	Recommended
5.	Camelids	Natural range 1/	Meat, wools, hides	Recommended
6.	Bovines, Dairy	Roughage, residues	Milk	Recommended
7.	Bovines, Draft	Roughage, residues	Work	Recommended
8.	Bovines	Preferably untillable grassland	Milk, blood	Acceptable
9.	Steers	Natural range 1/	Meat, hides	Acceptable
10.	Poultry	Grain, feed	Meat, eggs	To be avoided
11.	Swine	Grain, feed	Meat, hides	To be avoided
12.	Steers	Fertilized pasture	Meat, hides	To be avoided
13.	Steers	Feedlot or grain	Meat, hides	To be avoided
14.	Steers	Pasture from moist forest 2/	Meat, hides	To be avoided

1/ Untillable tracts, or tillable tracts in countries where all domestic food crop requirements have been adequately met.

2/ Tropical moist forest destruction is causing worldwide concern (Chapter 12) and much of it is due to the establishment of beef pastures at the forest's expense. These pastures create little employment and are almost always unsustainable, being abandoned within a few years.

The above ranking (Table 13.6) of necessity generalizes many distinctions which may influence project design. For example, beef production on untillable grasslands and in agro-forestry is often environmentally acceptable (Plucknett, 1979), but much less so for projects requiring high-energy inputs. This is one reason why livestock projects with a tsetse control component (generally aircraft and petroleum-based biocides) are environmentally more questionable than those in tsetse-free areas (Chapter 19). Livestock production can be less of an environmental problem to the extent that fertilizer and biogas are produced from the manure.

Milk and Dairy Products: There is clearly a significant role for production of milk and dairy products. Milk can be a valuable source of calcium and protein for childen in particular. In India's AMUL project 240,000 farmers, most of whom have only one cow or buffalo produce 700,000 liters of milk per day (less than three liters per farm) (Schaefer-Kehnert, 1981). Multi-purpose water buffalo in Asia are especially valuable (NAS, 1982). In Asia and elsewhere in the wetter tropics, water buffalo fill a unique role in rice culture as work animals, milk producers (especially the Indian breed), and converters of the relatively coarse, high silica-content rice straw, (their normal food), along with miscellaneous roadside vegetation. Since buffalo are well-adapted to rural conditions, they require no special care. There is more protein and calories in buffalo milk than in cow's milk, although the yield is less (Table 13.7). Buffalo are of great value as draft animals (Ward, et al. 1980). They can be partly sustained on naturally produced fodder on non-arable or fallow lands. Both bulls and cull cows can also be used as draft animals, becoming useful by-products of dairy herds.

Manure and Recycling: The contribution of manure for fertilizer, fuel, or construction from recycled crop and domestic residues that otherwise would be wasted further improves the environmental equation. Again, buffalo can recycle coarser materials (such as rice straw, and water hyacinths) than cattle commonly do. Neither animal can dispose of the 90-95 percent water content fast enough to be able to subsist entirely on water hyacinth, but reduction of water content to about 70 percent by means of a simple screw press raises the nutrient concentration to that of fresh alfalfa -- a nutrient level upon which both animals can thrive. A most useful role of swine (Chapter 14), and to some extent bovines, is in individual households where the animals can be stabled or kept under the dwelling to process domestic wastes. Fodder cut by hand elsewhere and walked to the usually solitary cow can eliminate overgrazing, and trampling of crops and soil compaction. This is particularly important in erosion-prone, high-rainfall areas with steep slopes, as in Java.

Table 13.7: Composition of Milk per 100 grams (East Asia)

Type of Milk	Food Energy (calories)	Protein	Fat	Carbo-hydrate	Calcium (milli-	Phosphorus (milli-	Iron (milli-
					grams		
Cow, whole fluid	63	3.1	3.5	5.0	114	102	0.1
Buffalo, whole fluid	115	5.2	8.7	4.3	210	101	0.1
Goat, whole fluid	65	3.4	3.8	4.5	142	118	0.1
Human, whole fluid	62	1.5	3.2	7.0	34	20	0.2
Horse, whole fluid	40	2.1	1.1	5.4	-	-	-

Source: FAO, 1972. (cf. Watt, et al., 1963).

Desertification and Livestock Projects: The world's arid and semi-arid zones contain more than one third of its sheep, over half its cattle, and two-thirds of its goats. Cattle production may be highly appropriate for areas where aridity prevents cropping. However, cattle and other livestock can increase the susceptibility of such areas to the process of desertification.

Along the southern fringe of the Sahara, 650,000 km^2 have been lost to desert--much of it caused by humans--over the last 50 years. Worldwide desertification appears due in great measure to increased pressure from grazing, cropping and fuelwood gathering. Livestock production exceeding the carrying capacity of an area leads to the widespread--and increasing--syndrome of overgrazing, involving deterioration of grazing land, particularly around watering holes; loss of annual and then perennial vegetation; compaction of soils; and subsequent increases in erosion and flooding. Changing political, social and economic factors also contribute to this deterioration; limits on the mobility of nomads by government controls, establishment of boundary and invasion by agriculturalists have altered the traditional balance between nomadic pastoralism and the environment. With space curtailed, nomadic pastoralism becomes more damaging to the range. Commercial ranching is even less adaptive to changes in carrying capacity than pastoralism, since animal movements are more constrained and the ranch specializes in one kind of animal, selected usually for the value of its product rather than its suitability to the environment. The UN Conference on Desertification (1977) reported that:

"Capitalization and technical improvements tend to buffer commercial ranching against the immediate consequences of overgrazing .... high prices for (and maximization of) its products will yield (short term) cash returns which may further delay a proper response to degradation of pastures. Unlike the more traditional systems, commercial ranching (using) heavy machinery for construction and road building,... can disturb the environment, producing localized degradation. The generally greater capacity for ecological manipulation in these technically advanced systems may have drastic feedback consequences."

In summary, where livestock production is promoted in semi-arid and arid regions susceptible to desertification, extra care must be taken in livestock management and prevention of overgrazing. Proposed changes in traditional livestock systems should be assessed to determine whether such changes will contribute to desertification as well (Grainger, 1982). Game ranching can be environmentally much more appropriate and less damaging in semi-arid lands than cattle ranching (Chapter 19). There is no need for "forests (to) come before civilizations; (nor for) deserts to follow them" as Chateaubriand observed.

References

Crotty, R.O. 1980. Cattle economics and development. Farnham Royal, UK, Commonwealth Agric. Bureau, 253 p.

Doll, E.C. and Mott, G.O. (eds.) 1975. Tropical forages in livestock production systems. Madison, Wisconsin, American Society of Agronomy, Special Publication 24:104 p.

Ecologist, 1976. Should an ecological society be a vegetarian one? The Ecologist 6(10):356-384.

Eckholm, E. and Brown, L.R. 1977. Spreading deserts: The hand of man. Washington, D.C., Worldwatch Paper 13: 40 p.

FAO, 1976. Conservation in arid and semi-arid zones. Rome, FAO Conservation Guide 3: 137 p.

FAO, 1972. Food composition table for use in East Africa. Rome, FAO, 334 p.

Feder, E. 1980. The odious competition between man and animal over agricultural resources in the underdeveloped countries. Binghampton, N.Y., Review II:3.

Grainger, A. 1982. Desertification: how people make deserts, how people can stop and why they don't. London, Earthscan, 94 p.

Green, M. B. 1978. Eating oil: energy use in food production. Boulder, Colorado, Westview Press, 205 p.

Hecht, S. B. 1981. Cattle ranching in the Amazon: analysis of a development strategy. Berkeley, California, University of California dissertation. (Geography and Forestry):450 p.

Hegsted, D.M. 1979. Shortening the food chain by replacing nutrients of animal origin (24-27) in Animal disease prevention in developing countries. Washington, D.C., Pan Amer. Health Org. Sci. Publ. 380:58 p.

Lappe, F.M. 1978. Diet for a small planet. N.Y., Ballantine, 410 p.

Leach, G. 1976. Energy and food production. Guildford (UK), IPC Sci. Tech. Press 137 p.

Makhijani, A. and Poole, A. 1975. Energy and agriculture in the third world. Cambridge, Massachusetts, Ballinger, 168 p.

Minish, G.L. 1979. Beef production and management. Reston, Virginia, Reston Pub. Co., 416 p.

Morley, F.H.W. (ed.) 1981. Grazing animals. Amsterdam, Elsevier
    411 p.

Murrary, R.M. and Entwistle, K.W. (eds.) 1978. Beef cattle
    production in the tropics: course notes and selected references
    for short course on tropical cattle production held at the
    Department of Tropical Veterinary Science, Townsville,
    Australia, University of North Queensland, 373 p.

NAS, 1982. Diet, nutrition, and cancer. Washington, D.C., National
    Academy of Sciences, 446 p.

NAS, 1982. The water buffalo: new prospects for an underutilized
    animal. Washington, D.C., National Academy of Sciences, 118 p.

Nations, J. D. 1978. Indigenous agroecosystems and the export beef
    cattle industry in tropical Latin America. New Delhi, 10th
    International Congress of Anthropological and Ethnological
    Sciences, 23 p.

Odum, H. T. 1971. Environment, power and society. New York, Wiley,
    331 p.

Odum, H. T. and Odum, E.C. 1976. Energy basis for man and nature.
    New York, McGraw-Hill, 296 p.

Pimentel, D. et al. 1973. Food production and the energy crisis.
    Science 182:443-449.

Pimentel, D. 1975. Energy and land constraints in food-protein
    production. Science 190:754-761.

Pimentel, D. 1976. The energy crisis: its impact on agriculture.
    Milan: Encyclopedia della Scienza e della Tecnica,
    Mondadori:251-266 p.

Pimentel, D. 1977. Energy resources and land constraints in food
    production. New York, Anrals NY Acad. Sciences 300:26-32.

Pimentel, D. et al., 1979. A cost benefit analysis of pesticide use
    in U.S. food production (97-149) in Sheets T.J. and Pimentel, D.
    (eds.) Pesticides: their contemporary roles in agriculture,
    health and the environment. Clifton, New Jersey, Humana Press,
    186 p.

Pimentel, D. et al. 1980. The potential for grass-fed livestock:
    resource constraints. Science 207:843-848.

Pimentel, D. and Terhune, E.C. 1977. Energy and food. Palo Alto,
    Annual Review of Energy 2:171-195.

Plucknett, D.L. 1979. Managing pasture and cattle under coconuts.
    Boulder, Colorado, Westview Press, 364 p.

Regan, T. 1982. All that dwell therein: animal rights and
    environmental ethics. Berkeley, California Univ. Press, 249 p.

Romanini, C. 1978. Agricultura tropical en tierras ganaderas: alternatives viables. Mexico, D.F., Centro Ecodesarrollo:103 p.

Schaefer-Kehnert, W. 1981. Appraisal and finance of intensive animal production schemes. Washington, D.C., World Bank, EDI Course Note Series CN-75:21 p.

Simpson, J.R. and Farris, D.E. 1982. The world's beef business. Ames, Iowa, Iowa State Univ. Press, 334 p.

Singer, P. 1975. Animal liberation: a new ethics for our treatment of animals. New York, Avon Discus, 297 p.

Smith, A. J. (ed.) 1976. Beef cattle production in developing countries. Edinburgh, Center for Tropical Veterinary Medicine, 487 p.

United Nations Conference on Desertification, 1977. Desertification: an overview--processes and causes of Desertification. Nairobi, Kenya, 79 p.

UNESCO/Man and the Biosphere Programme. 1974. Impact of human activities and land use practices on grazing lands: savanna, grassland (from temperate to arid areas). Hurley (UK), MAB Report Series No. 25: 97 p.

UNESCO, 1975. Obstacles to the development of arid and semi-arid zones. Paris, UNESCO, Nature and Resources 11 (4):2-11.

Ward, G.M. 1983. Energy impacts upon future livestock production. Boulder, Colorado, Westview, 250 p.

Ward, G.M., Sutherland, T.M., and Sutherland, J.M. 1980. Animals as an energy source in third world agriculture. Science 208:570-574.

Watt, B.K. et al., 1963. Composition of foods. Washington, D.C., USDA, Agr. Research Serv., Handbook 8:190 p.

Winrock Report, 1977. Potential of the world's forages for ruminant animal production. Morrilton, Arkansas. Winrock International Livestock Research and Training Center, 90 p.

Winrock Report, 1978. The role of ruminants in support of man. Morrilton, Arkansas. Winrock International Livestock Research and Training Center, 136 p.

World Bank, 1978. Slaughterhouses II: Plant design and equipment. Washington, D.C., Office of Environmental Affairs, World Bank, 6 p.

World Bank, 1979. Slaughterhouses I: Industrial waste disposal. Washington, D.C. Office of Environmental Affairs, World Bank, 5 p.

14
Small Livestock

Small livestock are defined here as sheep, goats, pigs, rabbits, agoutis (rabbit-sized Neotropical rodents), guinea pigs, capybaras (giant, semi-aquatic rodents), poultry, and other less commonly raised fowl and small mammals, as opposed to the large livestock such as cattle and water buffalo) discussed in Chapter 13. The raising of horses for chevaline appears insignificant compared with worldwide cattle production, so is not mentioned further. Production of small livestock, carried out as an adjunct to other agricultural activities, can be designed to exploit agro-ecological niches that would otherwise remain unutilized. Small ruminants can subsist on forage which humans cannot directly consume; pigs and fowl scavenge for household and agricultural wastes. These animals can act as a reserve at times when grain is available in excess of human needs. Since they are small, they can be eaten or sold more readily when grain is unavailable. Ducks and geese may also obtain nourishment from waterweeds and aquatic organisms. More unconventional livestock, such as agoutis, capybaras, and rabbits, offer considerable potential as small-scale and localized sources of protein and as generators of employment.

With the exception of sheep and goats, small livestock are not generally associated with environmental degradation and can largely be regarded as environmentally beneficial (Bishop, 1982). Small livestock production is usually intensive; most species do not require the creation of pasture and thus do not compete with crop production for arable land. Small livestock are even more adept than large ruminants at gleaning nourishment from land unsuited to other forms of agriculture. Furthermore, the plant protein-to-animal protein conversion rates of small livestock are more efficient than those of large livestock (Table 13-2).

Small livestock act as local, decentralized sources of milk, eggs, and meat, thereby promoting household self-sufficiency, increasing protein intakes, and reducing the need for refrigeration, processing, transportation and other energy-intensive activities that the rural poor often cannot afford. Surplus animal products such as feathers, hides, and wool can be sold. Small livestock recycle household and agricultural wastes, while their manure is generally recycled near human dwellings in vegetable gardens and fishponds, and can also be used for biogas production.

As noted in Chapter 13, a major concern about livestock production is the competition between the use of grain for feed versus food. While small livestock convert grain to meat more efficiently than large livestock, grain-fed livestock production of any sort wastes energy (and nutrients) when compared with the direct human consumption of grain.

Goats and Sheep: While goats are frequently blamed for erosion, deforestation, and some range degradation, it is not always clear whether their activities initiate these problems, or if goats succeed because no other species is capable of subsisting on already degraded land. Indeed, goats can play a role in weed control. Nevertheless, in areas prone to erosion or desertification, considerable care is needed in goat and sheep husbandry.

Goats have several important advantages over cattle. Their smaller size makes them useful to poor rural families, for whom owning a single cow would be uneconomic or impracticable. Goat meat, chevon, has a much higher protein content than beef or mutton. Goat milk is also higher in protein and minerals than cow's milk, hence valuable for humans.

Sheep which graze on unimproved pasture are maintained in the tropics more for meat production than for wool, although they also supply skins, milk, manure and hair. Grazing on non-fallow land suited for crop production represents an inefficient form of land use, if local food demands are not fully met. More intensive production of sheep is possible; for example, the Kikuyu people in Kenya stall-feed sheep with succulent and nutritious fodder, such as sweet potato vine and groundnut haulm. Bran of wheat, rice, and sorghum, as well as other agricultural by-products, can also be stall-fed to sheep and other ruminants.

Other advantages in sheep and goat production include high reproductive efficiency; higher energetic efficiency of milk production (goats exceed any other stock animal in this regard); the ability of these animals, on account of their small size, to utilize marginal and small plots of land more efficiently than cattle; and small carcasses, which are easier to handle and can be consumed in short periods of time, an important consideration in regions lacking effective preservation methods or refrigeration.

Poultry: These include chickens ducks, geese, quail, guinea fowl, pigeons, turkeys, and other domesticated fowl. As long as poultry feed is limited to scavenging consumption of and domestic wastes, production will be low in quantity but free from the conflict between human and poultry food needs. Increased production demands a regular supply of nutritious food, which inevitably leads to food competition between humans and the domesticated poultry.

Grain feeding of poultry is wasteful of energy and nutrients, except where there is grain to spare after human needs have been satisfied. Poultry are more efficient than most livestock as converters of vegetable feed into animal protein. Poultry eggs rank with cow's milk as the most economically produced animal protein, and poultry meat ranks above other meats in this respect also. There are no major environmental disadvantages associated with poultry production, except on a relatively large scale.

Ducks are raised for both eggs and meat. Under intensive husbandry, both Indian Runner Ducks and chickens have been known to produce as many as 300 eggs per year. The most significant environmental aspects of duck and goose production are that these birds exploit a niche that otherwise would be unutilized by humans, and that they increase the productivity of other niches. This is best illustrated in the case of waterfowl on fish ponds and rice paddies. The birds eat aquatic organisms, including pests. They also fertilize the water or paddy with nitrogen-rich droppings, and thereby increase yields of both fish and rice. Moreover, the pond can be drained intermittently and the bottom soil recycled as fertilizer. Waterfowl and fish will also consume agricultural and household wastes.

Pigs: Approximately 20 percent of world pig production occurs in the tropics. The number of humans per pig varies from 60 in Africa to about two in South America, where the pig population is one of the densest in the world. The Moslem prohibition against eating pork means pigs are not raised in some countries. Although the primary purpose of pig farming is meat production, one large pig (maintained at bacon weight) produces one ton of manure per year, which can be treated either as a nuisance or as a resource.

Since pigs are omnivorous, they can compete with humans for food. However, since pigs are also useful and efficient recyclers of the by-products of human food and of other wastes, such competition generally is less significant than with grain-fed cattle. In developing countries, pigs are more numerous where large quantities of by-products or offal are available. Waste foods, such as rice bran and the peelings of roots such as taro and cassava, are usefully converted by these animals. Better management to increase productivity includes the fencing in of pigs, in which case it is important to consider the transport costs of feed such as root peelings, village offal, and waste fruits and vegetables. Intensive pig production is socially more desirable since it can assist small farmers more than intensive beef production does.

Potential increases in pig production near towns occur are most promising where butchers have slaughter offals, where breweries have by-products, and where domestic garbage is always abundant. (In the latter case, disease transmission needs attention.) In the rice-growing areas of Southeast Asia, rice milling by-products can support pig production. In the same region, there is a tradition of pig farming in association with fish-pond culture, because effluent from piggeries is conducted into the fish-ponds, thus stimulating the growth of micro-organisms on which the fish feed. If tropical fish (e.g., _Tilapia mossambica_) are cultivated in a pond, and aquatic plants (such as _Ipomoea reptans_) are grown on its surface, there exists both a natural outlet for effluent from the piggeries and a source of high-quality fish protein and green fodder for feeding the pigs. Piggery waste also can be used to generate biogas for household cooking, with a valuable nutrient-rich sludge as a by-product. Recognizing the benefits of small livestock, it is noted in relation to one project in the Philippines noted:

> "No adverse impact on the environment is expected. Effluent from large piggeries is typically placed in sewage treatment lagoons. The small piggeries' waste materials are often used to generate methane gas for household cooking. Alternatively, the waste is used in food gardens and rice fields. All droppings from poultry are dried and utilized as fertilizer for vegetable production or in fish-ponds."

In conclusion, the use of waste and non-conventional food sources for pigs is environmentally efficient. Feeding grain and other foodstuffs fit for human consumption to pigs is less efficient. Although more efficient than cattle in converting grain protein to meat protein, pigs rank behind chicken and other poultry in this conversion.

References

Adams, R., Adams, M., Willens, A., and Willens, A. 1978. Dry lands: man and plants. London, Architectural Press, 158 p.

AID, 1983. Complete handbook on backyard and commercial rabbit production. Manila, CARE/AID/Peace Corps:93 p.

Almazan, O. 197/. Sugar industry products as a source of animal feed in the tropics. ICIDCA V. 11(2-3):32-54.

Bishop, J.P. 1982a. Tropical forest sheep on legume forage/fuelwood fallows. Washington, D.C., USAID (ms):11 p.

Bishop, J.P. 1982b. The dynamics of shifting cultivation, rural poor, cattle complex on marginal lands in the humid tropics. Washington, D.C., USAID (ms):19 p.

Chenost, M. 197/. Utilization of waste products in animal feeding. (465-473) in Residue Utilization, Management of Agricultural and Agro-Industrial Residues. Rome, FAO: 2 vols.

Devendra, C. 1979. Pig production in the tropics. Oxford University Press, 172 p.

FAO, 1977a. China: recycling of organic wastes in agriculture. Rome, FAO Soils Bulletin 40:107 p.

FAO, 1977b. New feed resources. Rome, FAO Animal Production and Health Paper 4:300 p.

French, M.H. 1970. Observations on the goat. Rome, FAO Agricultural Studies 80:704 p.

Guzman, H.R. de, and Lee, N.S. 1978. Integration of backyard (small-scale) dairy beef farming with cropping systems and feed grain substitutes for cattle. Taipei, Taiwan. Food and Fertilizer Technology Center, Asian and Pacific Council, Extension Bulletin 110:29 p.

Holst, P.J. 1980. The use of goats in grazing systems and their place in weed control. Austr. Soc. Animal Prod. 13:188-191.

Land, T. 1978. Opening up tropical Africa for cattle production. Africa (UK) 86:74-75.

McDowell, R.E. 1977a. Are US animal scientists prepared to help small farmers in developing countries? Ithaca, New York, Cornell International Agriculture Mimeograph 58:25 p.

McDowell, R.E. 1977b. State of the dairy industry in warm climates of less developed countries. Ithaca, New York, Cornell International Agriculture Mimeograph 57:70 p.

McDowell, R.E. and Bove, L. 1977. The goat as a producer of meat. Ithaca, New York, Cornell International Agriculture Mimeo. 56:40 p.

McDowell, R.E. and Sands, M. 1979. A world bibliography on goats. Ithaca, New York, Cornell International Agriculture Mimeo. 70:112 p.

McGarry, M.G. and Stainforth, J. 1978. Compost, fertilizer and biogas production from human and farm wastes in the People's Republic of China. Ottawa, International Development Research Center, 94 p.

National Science Foundation, 1976. Increased productivity from animal agriculture. Ithaca, New York, Cornell University, 102 p.

Oltenacu, E.A., Martinez, A., Glimp, H.A., and Fitzhugh, H.A. (eds.) 1976. The role of sheep and goats in agricultural development. Washington, D.C., USAID Technical Assistance/ Agriculture:41 p.

Overcash, M. R. et al. 1983. Livestock waste management. Boca Raton, Fla., CRC Press, 2 vols.

Ranjhan, S.K. 1978. Use of agro-industrial by-products in feeding ruminants in India. Rome, FAO, World Animal Health 28:31-37.

Rockefeller Foundation, 1975. The role of animals in the world food situation. New York, Rockefeller Foundation, 101 p.

Shelef, G., Moraine, R. and Oran, G. 1978. Animal feed proteins and water for irrigation from algal ponds (217-228) in Water pollution control in developing countries. Proc. Intern. Conf. Bangkok, Thailand. Feb. 1978. 2 Vols:1027 p.

Taiganides, E.P. 1978. Energy and useful by-product recovery from animal wastes (315-323) (in above).

Voorthuizen, E.G. van. 1978. Global desertification and range management: an appraisal. J. Range Management 31(5):378-380.

Werner, T. 1968. Poultry keeping in tropical areas. Rome, FAO:59 p.

Williamson, G. and Payne, W.J.A. 1959. An introduction to animal husbandry in the tropics. London, Longmans Green, 447 p.

15
Freshwater Fisheries

Fish can provide protein at half to two-thirds the economic cost of mammalian protein. Environmentally, freshwater fishery projects are among the most desirable for vertebrate protein production. However, environmentally unsound agricultural, industrial, and urban development causes major damage to fisheries worldwide. These activities, with their competing water demands, are prime deterrants to the integration of fish farming into rural areas.

Aquaculture is the small-scale, labor-intensive propagation and culture of aquatic organisms under human control. When fully integrated into rural development schemes, aquaculture provides plentiful, reasonably priced, fresh, and locally available protein-rich food. It also generates much-needed employment. The biological efficiency of aquaculture, in terms of feed conversion and fertilizer response, is higher than that of most land-based livestock. Catfish production, for example, is similar in such efficiency to pork production, but less efficient than broiler chicken production.

Aquaculture is carried out in brackish coastal areas, irrigation networks, rice paddies, man-made and natural ponds, cages, tanks, pens, waste systems, and lagoons. When integrated with agriculture, aquaculture not only improves soils, but also adds protein to systems already productive of grain and other crops. In much of Asia, the daily protein from aquatic organisms harvested from rice fields and small ponds is on a par with the daily milk production in temperate zone rural areas. When linked to waste disposal systems, aquaculture elevates animal and human wastes from an expensive disposal and health problem to a valuable resource. Furthermore, high energy costs place aquaculture in an advantageous position as a source of food protein; the energy resources necessary to improve low-quality land for agricultural purposes are almost always much higher than if the same land is used for aquaculture. However, feeding grain to any fish "crop" suffers from the same high energy and environmental costs as do all feedlot systems.

Besides aquaculture, lake and river fisheries projects also provide employment and high protein yields with minimal adverse environmental impact, while promoting such environmental benefits as control of water weeds and decreased pressure on productive land. Fishery projects are vulnerable to pollution, civil works such as hydroelectric projects, irrigation schemes involving water diversion and sedimentation, and hydrological changes brought about by deforestation. Mismanagement, such as by overfishing, of what should be a sustained-yield system can easily destroy the fishery.

Impact of Fishery Projects on Water, Soils, and Wildlife: The culture of fish in waste-water treatment lagoon systems generally results in effluents of higher quality. Some fish clean the water by eating algae, bacteria, or fungi. Furthermore, well-managed waste-water aquaculture systems provide suitable environments for rearing fish for human consumption; waste waters contain low-cost protein sources for fish foods. However, fish are bio-accumulators and cannot be consumed by humans if heavy metals or persistent biocides are present.

When fully integrated with agriculture and animal production, aquaculture effectively converts farm wastes (such as manure, grass clippings, and crop and domestic residues) into protein. Following pond drainage or dredging, nutrient-rich sediments accumulating in ponds can be recycled and returned to soils as fertilizer. In some low-lying coastal areas, the digging of fish ponds lowers the water table and, where the pond has an outlet drain, corrects soil waterlogging.

Exotic fish species introduced for cultivation can wreak havoc in new environments. Such species should not be introduced until it is fully determined that there is no acceptable indigenous species, and only if the ecological dangers that could ensue from exotic introductions have been carefully considered. For example, the introduction of carp (Cyprinus carpio) into certain regions of Latin America has created serious problems. Its feeding habits have led to muddied water and damaged vegetation essential to the reproduction of indigenous species. Although carp is eaten by carnivorous fish in Brazil, there are areas in which natural predation on carp is not sufficient to prevent overpopulation, resulting in the destruction of habitat for local species (FAO, 1974).

In Southeast Asia, brackish water fishponds in estuarine swamp have been created at the expense of mangroves and their associated wildlife. The conversion of swamplands, especially those containing large lakes, into fishponds and fields can pose a threat to water quality, as this reduces the natural filtration of water (see discussion of mangroves, Chapter 17). Furthermore, mangroves and swamps in their natural state are important breeding grounds for wild birds, fish, molluscs, and arthropods. In China, large introduced clumps of grass anchored in lakes serve the ecological role of marsh lands and swamps for a few species of animals that have been affected by the swamps that were flooded, drained or otherwise altered for fish production.

Disease Vector and Water Weed Control through Aquaculture: When cultivated in irrigation ditches and canals, selected fish, such as topminnows (Gambusia) and guppies can help control disease vectors, especially mosquitoes, thereby reducing the need for biocides. There is some evidence that the Shellcracker Fish (Lepomis microlophus) may be of some value in controlling the snail vectors of schistosomiasis in ponds. Although real success has yet to be achieved in controlling schistosomiasis with fish, it is a promising avenue for research.

Herbivorous fish have been successful in controlling the proliferation of water weeds. The Chinese Grass Carp or White Amur (Ctenopharyngodon idella) eats vast quantities of water milfoil (Hydrilla) and Ceratophyllum in Gatun Lake of the Panama Canal. Nile Tilapia (Tilapia nilotica) and Congo Tilapia (T. melanopleura) are known to eat specific deleterious plants. The Silver Carp (Hypopthalmichthys molitrix) feeds exclusively on planktonic algae and could be used to control algal blooms. Biological control of water weeds reduces herbicide use. The large aquatic mammal, the manatee (Trichechus), successfully controls water weed in those areas where it can be protected from overhunting.

Environmental Problems that Harm Fishery Production and Potential: Deforestation (Chapter 12) causes watershed changes and the sedimentation of rivers and lakes (natural and man-made), thereby harming fish populations. Large civil engineering works such as hydroelectric dams and irrigation schemes also create major impacts on fish populations. Fish migrations are often impeded or even prevented, habitats are modified or eliminated, and water quality may deteriorate in reservoirs. In the water above dams, riverine populations either adapt to the lake conditions or disappear. In certain cases, dam projects provide "ladders" or other fish passage facilities to allow fish migration, a provision that has had some limited success with fish capable of surviving in large reservoirs.

Chemicals and heavy metals either poison fish, shellfish, crustaceans, and other organisms directly or accumulate in their tissues, thus rendering them unsafe for human or animal consumption. The hardshell clam (Mercenaria), for example, filters eleven liters of water every hour, thereby having the potential to accumulate substantial quantities of toxic wastes. Concern over the impact of industrial effluent in one aquaculture project resulted in exclusion from the project area of those industries deemed most harmful to aquatic life. Project design also included monitoring of biocide levels in local water.

Decreases in river, delta, and marine fish populations may be related to dam construction upstream. For example, declining nutrient loads in the Nile as a result of the Aswan High Dam have virtually eliminated Mediterranean sardine fishery, which formerly produced 18,000 tons per year (George, 1972). Reduced nutrient load and water plant populations also increase river velocity, which in turn leads to scouring and erosion of the river bed and bank, which

damages breeding habitat and affects fisheries in other ways.
Similarly, flood control measures can also greatly affect the
productivity of freshwater fisheries and fish-pond development.

Drainage and reclamation projects which prevent the mixing
of salt and fresh water in estuarine areas can affect the breeding
and feeding activities of numerous fish and mollusc species. In Sri
Lanka, aquatic life suffered and fishermen experienced reduced
catches because a project impeded the free mixing of sea and fresh
water and limited access by fish to a frequently flooded marshy
area. The fishermen's losses were substantial. These losses are
all the more important to a nation like Sri Lanka that relies
heavily on fishing for the protein supply of its population and for
exports. Improved project design could have mitigated the project's
negative impact on marine life. Any interference with the free
mixing of salt, brackish, and fresh water may lead to costly
environmental deterioration, including reduced aquatic life.

Environmental Aspects of Fishery Project Design: Whether
in aquaculture or river and lake fishery projects, fish and other
aquatic organisms should be regarded as a renewable resource that
can be sustained in perpetuity--but only if properly managed. In
the design and justification of national fisheries programs, it is
prudent to consider environmental concerns, natural resource
conservation, and the interrelationships between fisheries projects
and other sectors of the economy. Environmental aspects of
fisheries projects include watershed and coastal zone management,
artificial stocking of fish in rivers and lakes, improvement in
present resource management techniques (such as estimating maximum
sustainable yield), and use of surveillance techniques to prevent
overexploitation. Water quality monitoring and surveillance of fish
processing facilities are important to detect contamination.
Industrial projects, urban projects (e.g., sewage systems), and
agricultural projects (e.g., irrigation systems) with potential for
pollution should have their potential impact on fisheries assessed
and reduced through appropriate mitigatory measures.

References

Ackermann, W. C. et al. (eds.) 1973. Man-made lakes: their
problems and environmental effects. Washington, D.C., American
Geophysical Union, 847 p.

Anon. 1972. A discussion of freshwater and estuarine studies on
the effect of industry. (Organized by Sir Frederick Russell,
F.R.S., and H.C. Gilson). Proc. Royal Soc. Lond. R.
180:365-536.

Baur, R.J., Buck, D.H. and Rose, C.R. 1978. Utilization of swine
manure in a polyculture of Asian and North American Fishes.
Transactions of the American Fisheries Society, 107 (1): 216-222.

Bell, F.W. 1976. Agriculture for developing countries; a
feasibility study. Cambridge, Mass. Ballinger Pub. Co. 226 p.

Buck, D.H. 1977. The integration of aquaculture with agriculture.
Fisheries 2(6):11-12, 14-16.

Dill, W.A. and T.V.R. Pillay. 1968. Scientific basis for the
conservation of non-oceanic living aquatic resources. Rome, FAO
Fish. Tech. Pap:829 p.

Edmondson, W.T. (ed.) 1971. Methods for estimation of secondary
productivity in fresh waters. Oxford, Blackwell, IBP Handbook
17:358 p.

FAO, 1974. Aquaculture in Latin America. Rome, FAO Fisheries
Report No. 159: 44 p.

FAO, 1975. Symposium on aquaculture in Africa. Rome, FAO,
Committee for Inland Fisheries of Africa (CIFA). Technical Paper
No. 4: 36 p.

FAO, 1976a. Biological effects of pollutants: bioassays with
aquatic organisms in relation to pollution problems. Rome, FAO
Fisheries Report 187: 28 p.

FAO, 1976b. Report of the FAO technical conference on aquaculture.
Rome, FAO Fisheries Report 188:93 p.

FAO, 1977. Freshwater fisheries and aquaculture in China. Rome,
FAO Fisheries Technical Paper 168: 84 p.

George, C.J. 1972. The role of the Aswan High Dam in changing the
fisheries of the Southeastern Mediterranean (159-178) in Farvar,
T. and Milton, J. (eds.) The careless technology. New York,
Garden Press, 1030 p.

Golterman, H.L. (ed.) 1969. Methods for chemical analysis of fresh
    waters. Oxford, Blackwell, IBP Handbook 8: 172 p.

Kajak, A. and Hillbricht-Ilkowska, A. (eds.). 1972. Productivity
    problems of fresh water. Warsaw and Krakow, IBP/Unesco
    symposium:918 p.

Lachner, E.A., Robins, C.P. and W.R. Courtnay, 1970. Exotic fishes
    and other aquatic organisms introduced into North America.
    Smithsonian Contr. Zool. 59:29 p.

Lauff, G.H. (ed.). 1967. Estuaries. Washington, D.C., Am. Ass.
    Advmt. Sci. 83:757 p.

Likens, G.E. (ed.). 1972. Nutrients and eutrophication: the
    limiting-nutrient controversy. Amer. Soc. Limnol. Oceanogr.
    Spec. Symp. Vol. 1.

Loftus, K.H. and Regier, H.A. (eds.). 1972. Salmonid communities
    in oligotrophic lakes: effects of three major cultural stresses
    - fisheries exploitation, nutrient loading, and exotic fish
    species. J. Fish. Res. Bd. Canada 29:611-986.

Lowe-McConnell, R.H. (ed.). 1966. Man-made lakes. London,
    Academic Press, 218 p.

Lowe-McConnell, R.H. 1975. Fish communities in tropical fresh
    waters; their distribution, ecology and evolution. London,
    Longmans Green, 337 p.

Luther, H. and J. Rzoska. 1971. Project Aqua: a source book of
    inland waters proposed for conservation. Oxford, Blackwell, IBP
    Handbook 21:239 p.

Man and the Biosphere Green Books. 1972 and 1974. Ecological
    effects of human activities on the value and resources of lakes,
    marshes, rivers, deltas, estuaries, and coastal zones. London,
    (Two Issues) Sept. 1972 and May 1974, 49 p. and 80 p.

McLarney, W.O. 1976. Aquaculture: toward an ecological approach:
    fish ponds on farms. (328-339) in Merrill, R. (ed.) Radical
    agriculture. New York, Harper and Row, 459 p.

Melchiorri-Santolini, U. (ed.). 1972. The role of detritus in
    aquatic ecosystems. IBP/Unesco Symposium, Mem. It. Ital.
    Idrobiol., (Supplement):540 p.

Milway, C.P. (ed.). 1970. Eutrophication in large lakes and
    impoundments. Paris, OECD:562 p.

Morales, H.L.1978. La Revolucion Azul? Aquaculture y
    Ecodesarrollo. Mexico D. F., Centro de Ecodesarrollo, Programa
    de las Naciones Unidas para el Medio Ambiente. Mexico D.F.
    Editorial Nueva Imagen:159 p.

National Academy of Sciences. 1969. Eutrophication: causes, consequences, correctives. Washington, D.C., National Academy of Sciences, 661 p.

Obeng, L. (ed.) 1969. Man-made lakes: the Accra Symposium. Accra, University of Ghana Press, 398 p.

Poppensiek, G.C. 1972. Aquaculture and agriculture: protein potentials. Amer. Vet. Med. Assoc. J. 161(11):1467-1475.

Prasad, R.R. 1973. Protein crisis and aquaculture. Indian Farming 23(6):5-38.

Ricker, W.E. (ed.). 1971. Methods for assessment of fish production in fresh waters. Oxford, Blackwell, IBP Handbook 3:348 p.

Rouzaud, P. 1973. L'aquaculture: source nouvelle de proteines alimentaires. Bull. Soc. Sci. Hyg. Aliment. 61(1):18-31.

Schraeder, G.L., 1978. Agriculturalists in fish farming. Dor, Israel. The Commercial Fish Farmer 4(6): 33 p.

Scientific Committee on Problems of The Environment, 1972. Man-made lakes as modified ecosystems. Paris, ICSU, SCOPE Report 2:76 p.

Seminar-Workshop-Proceedings on Artisan Fisheries Development and Aquaculture in Central America and Panama. 1976. San Jose, Costa Rica. International Center for Marine Resource Development (v.p.).

Stanley, B. 1978. Fish Farming: an account of the aquaculture research program supported by IDRC. Ottawa, IDRC, 40 p.

Tripathi, S.D. 1976. Role of fresh-water aquaculture in integrated rural development in India. Agric.Agroind. J. 9(12):31-32.

UNEP, 1980. Coastal ecosystems: a review. Nairobi, UNEP Report No. 4:65 p.

United Nations. 1970. Integrated river basin development. New York, United Nations Department of Economic and Social Affairs, 80 p.

Vollenweider, R.A. (ed.). 1969. A manual on methods for measuring primary production in aquatic environments. Oxford, Blackwell, IBP Handbook 12:213 p.

World Bank, 1979. The World Bank and the fishery sector in developing countries. Washington, D.C., World Bank, Economics and Policy Division, Agricultural and Rural Development Department, 115 p.

16
Marine Fisheries

As a result of today's highly sophisticated fish-harvesting technology, industrial fishing fleets are now increasingly able to upset the marine food chain, deplete fish populations, and overwhelm traditional fishing industries. These large fishing fleets frequently overexploit certain ecologically complex fish species (e.g., anchovy), while much of their catch is either exported to richer nations or is reduced to commercial fish products (e.g., fish meal) not directly consumed by humans. Worldwide, nearly one-third of all harvested fish is fed to livestock. Fish protein represents only about 5 percent of annual world human protein intake (Pimentel and Pimentel, 1979).

Several broad ecological observations are immediately apparent with regard to the world fish harvest. Productivity of some ocean fish stocks is declining due to overfishing, while the fish resources of freshwater bodies are either sub-optimally managed or are being depleted by mismanagement (e.g., pollution and engineering works). Freshwater aquaculture is labor intensive, and is often coupled with crop cultivation (e.g., rice, or use of pond sludge as fertilizer), or with waste disposal systems. Fish convert nuisance aquatic plants and insects into protein, while decreasing the need for biocides and potentially decreasing medical problems such as malaria (Chapter 15).

Marine fishery depletion results from the excessive exploitation of a small number of fish species. Of the approximately 20,000 fish species known, only 100 species make up 70 percent of the world catch. The Peruvian anchovy (Engraulis ringens) alone made up one fifth of the world's marine catch in 1971. For ecological and economic reasons, harvest of fish species declining in stock (or fish species which will decline in stock once the catch capacity of the country or area is increased) can be short-term only. Preferably, diversification of catch can be promoted, either through the introduction of new fishing technologies or through marketing assistance for previously "uncommercial" fish.

Approximately 70 percent of the world's marine fish harvest consists of pelagic species -- fish that reside in the upper strata of oceans and occupy the lower links of the aquatic food chain. Plankton-eating fish make up over 60 percent of the marine harvest, carnivorous fish comprise about 25 percent, and benthic (bottom-dwelling) species only 4 percent. The abundant resources of fish are those lower down in trophic level (food chain); however, overfishing of these species threatens the ecological stability of all the other trophic levels. Pelagic species, frequently associated with both high productivity and high total biomass, are largely short-lived species with high natural mortality and can support more intense harvests than can longer-lived species. The intensive technologies and commercial pressures which developed in response to this biological phenomenon are difficult to control and have contributed to the recent Icelandic "Fish War", the decline of the Peruvian anchovy, and the collapse of the sardine industry in California and the herring industry in British Columbia, the English Channel, and the North Sea.

While fishery projects can increase the ability of nations to exploit the ocean's resources, components of such projects also must be included to preserve the spawning and breeding grounds of marine organisms. Research results from West Bengal, for example, indicate that potential shrimp yields from 44,000 hectares of tidal mangrove swamps could amount to 1,000 tons of shrimp per year. Other studies show that more than 90 percent of marine species in a region are predictably found in mangrove swamps during one or more periods in their life cycle; for many species, it appears that the relationship is obligatory. Thus, the destruction of mangroves results in considerable decline of shrimp stock and other marine species. Investment in mangrove reforestation and protection would not only benefit certain marine populations but also would provide such functions as shore stabilization, flood control, and the sustained production of wood and other products for domestic and commercial uses (Chapter 17). Here a typical environmental aspect to be addressed in project design is careful siting of new harbors or other construction works to avoid destruction of breeding grounds. Similarly, other coastal developments unrelated to the fishing industry can pollute, drain, or otherwise destroy important breeding areas, even for some species caught far offshore. Dams on major rivers can substantially decrease nutrient flow into oceans and seas, thus decreasing dependent marine fish populations.

The blasting of coral reefs for harbor construction, agricultural lime, and cement production further degrades fisheries by eliminating the habitat of many valuable tropical species (Chapter 24). It also disrupts coastlines, making them more vulnerable to erosion.

Coastal lagoons and estuaries are among the world's most biologically productive natural ecosystems and provide vital breeding grounds for many marine fishery species. In terms of primary productivity, these ecosystems are estimated to be ten to fifteen times more productive than the waters of the continental shelf. In the coastal lagoons of Asia and West Africa, fishing yields have exceeded one ton of fish per hectare annually, which is significantly more than the protein output of prime grazing land.

Since lagoons and deltas are constantly enriched by alluvium, they are often "reclaimed" for agricultural development. Alternatively, they are dredged or filled for port facilities or other industrial development, or are badly polluted by urban or agricultural wastes. Unless special precautions are taken, however, such development can harm offshore fisheries. Some of the lagoons along the West African coast, for example, have been severely altered ecologically by the construction of port facilities, and fisheries resources have declined sharply.

The growth in domestic, agricultural, and industrial wastes concomitant with urbanization often accelerates eutrophication of coastal logoons sometimes resulting in major losses of fish, crustaceans, and molluscs. Since waste water from domestic refuse and sewage may not be fully purified by the lagoon ecosystem, it can also spread such diseases as viral hepatitis, cholera, and typhoid. Hydrocarbons and detergents in the waste water are partially broken down by bacteria, but heavy metals and biocides are often concentrated in food chains. The future of coastal lagoons, the rich fisheries they support, and local human health all depend importantly on the improved management of dumps and other urban waste disposal systems.

References

Anderson, J.I.W. 1975. The aquaculture revolution: cultivation of marine fish and shellfish for food. Proc. R. Soc. Lond, B. Biol. Sci. 191(1102):169-184.

Barnes, R.S.K. and Hughes, R.N. 1982. An introduction to marine ecology. Oxford, Blackwell, 240 p.

Clark, J. 1974. Coastal ecosystems: ecological considerations for management of the coastal zone. Washington, D.C. The Conservation Foundation, 197 p.

Freeman, P. 1974. Guidelines for policy, assessment and monitoring in tropical regions: Coastal zone pollution by oil and other contaminants. Washington, D.C. Office of International and Environmental Programs of the Smithsonian Institution, 67 p.

James, D. 1979. The international fisheries scene: changing directions and new priorities. Rome, FAO Fisheries Dept.

Lasserre, P. 1979. Coastal lagoons. Paris, UNESCO/MAB Bull. 15(4): 2-21.

Lauff, G.H. (ed.). 1967. Estuaries. Washington, D.C., Am. Ass. Advmt. Sci.:757 p.

Lowe-McConnell, R.H. 1977. Ecology of fishes in tropical waters. London, Arnold, 64 p.

MacNae, W. 1974. Mangrove forests and fisheries. Rome, FAO, IOFC/DEV/34/74:35 p.

McVey, J.P. (ed.). 1983. Handbook of mariculture. Boca Raton, Fla., CRC Press, 480 p.

Man and the Biosphere Green Books, 1972 and 1974. Ecological effects of human activities on the value and resources of lakes, marshes, rivers, deltas, estuaries, and coastal zones. London. Sept. 1972 and May 1974. 49 p. and 80 p.

Morales, H.L. 1978. La Revolucion Azul? Aquacultura y Ecodesarrollo. Centro de Ecodesarrollo, Programa de las Naciones Unidas para el Medio Ambiente. Mexico, D.F., Editorial Nueva Imagen, 159 p.

NAS, 1970. Waste management concepts for the coastal zone: requirements for research and investigation. Washington, D.C., National Academy of Sciences - National Academy of Engineering. 126 p.

Pearson, C. 1982. Environmental policies and their trade implications for developing countries, with special reference to fish and shellfish, fruit and vegetables. Geneva, UNCTAD/ST/MD/26:48 p.

Pimentel, D. and Pimentel, M. 1979. Food, energy and society. London, Arnold, 165 p.

Poppensiek, G.C. 1972. Aquaculture and mariculture: protein potentials. Amer. Vet. Med. Assoc. J. 161(11):1467-1475.

Ragotzkie, R.A. (ed.). 1983. Man and the marine environment. Boca Raton, Fla., CRC Press, 208 p.

Ryman, N. (ed.) 1981. Fish gene pools: Preservation of genetic resources in relation to wild fishstock. Stockholm, The Editorial Service FRN:111 p.

Shang, Y.C. 1973. Comparison of the economic potential of aquaculture, land animal husbandary and ocean fisheries: the case of Taiwan. Aquaculture 2(2):187-195.

World Bank. 1979. The World Bank and the fishery sector in developing countries. Washington, D.C., World Bank, Economics and Policy Division, Agricultural and Rural Development, 115 p.

17
Biocides in the Environment

The use of biocides for pest and disease control in crop production has played an important role in the past, and will continue to do so in the future to ensure adequate food supply for the growing world population. Pests have continually caused reduction in potential crop yields, and consequently have affected the world's potential food supply. A considerable percentage of potential crop yields is lost every year due to pests or diseases. Biocides help enormously to reduce these food losses. Estimates of world food losses due to diseases are summarized in Figure 17.1.

Figure 17.1:   Losses from Diseases in the World's Major Crops

Crop	Percent Loss	Production-MMT-1974		
		Actual	Potential	Loss
Wheat	9.1	360	396	36
Paddy Rice	8.9	323	354	31
Potatoes	21.8	294	376	82
Maize	9.4	293	323	30
Sweet Potatoes	5.0	134	141	7
Cassava	16.6	104	125	21
Millet and Sorghum	10.6	93	104	11
Soybeans	11.1	57	64	7
Bananas	23.0	36	47	11
Tomatoes	11.6	36	41	5
Peanuts	11.5	18	20	2

Source: Pimentel, 1978.

MMT = million metric tons.

While biocides play an important role in increasing yields, questions have been raised concerning their effects on human health and the environment. Chemical control of pests frequently involves the application of highly toxic substances. This chapter outlines some of the possible environmental and health consequences associated with the use of biocides.

The term "biocide", meaning chemical killer of life, as used here, has been deliberately chosen as a warning to remind the reader that the use of such substances usually brings undesired, as well as desired, effects. The generic term biocide includes herbicides, weedicides, arboricides, insecticides, pesticides, larvicides, fungicides, miticides, rodenticides, acaricides, nematocides, molluscicides, and related substances. "Pesticide", meaning killer of pests, is preferred by some people but is not used here since any so-called pesticide introduced into the ecosystem is unlikely to be target-specific. There are no species-specific biocides. The term pesticide suggests that pests can be distinguished from non-pests, that pesticides will not kill non-pests, and that pests are wholly and always undesirable (cf. Hardin, 1972). Rachel Carson exposed these fallacies in her classic work, Silent Spring, as long ago as 1962. All miticides, for example, are biocide-induced, since no mite was a crop pest before biocides were used. By 1980, nearly 400 species of insect and mite pests were known to be biocide-resistant.

Many of the early and now obsolescent generation of biocides--DDT, for example--were inexpensive and effective when first used, and as a result held out great hopes for benefitting humanity. Unfortunately, many (e.g., DDT, dieldrin, chlordane, heptachlor, BHC) were persistent, non-selective and damaging to non-target species. DDT, for example, could be called an avicide, since it is particularly effective in killing birds (Hardin, 1972). To the extent that modern biocides are more target-specific and less persistent in the environment, many deleterious side-effects will decrease as the obsolete biocides are phased out.

Global biocide use--500,000 tons of biocides per year recently in the United States alone--continues to increase significantly, particularly in developing nations. While there is extensive knowledge and ability to regulate biocides in the developed world, poisoning, pollution, misuse, overuse, and diminishing crop returns on heavy biocide investments are common in developing countries. For example, one incident of food contamination by alkyl-mercury biocide killed more than 500 people and hospitalized another 6,000 in Iraq in 1932. This type of calamity can result precisely because important precautions concerning biocide use are much less well known in developing countries.

While some developed countries have moved to regulate biocide exports, developing countries have become increasingly self-sufficient in biocide production. WHO estimates that eight developing nations already produce over one third of the world's DDT. Biocides are costly agricultural inputs; since 80 percent of synthetic biocides are manufactured from a petroleum base, costs (over US$12/kg in 1980) will continue to increase as the world's petroleum reserves become depleted. Furthermore, heavy and frequent biocide applications initially produce boom yields, but these usually taper off as pests and diseases develop resistance to biocides, as beneficial natural enemies are reduced or even annihilated, and as previously insignificant organisms assume pest roles.

Effects on Public Health: Public health problems in the tropics, in which mosquito related diseases alone afflict millions, make the use of biocides for disease vector control unavoidable in the short term. Ironically, the widespread use of biocides for agriculture renders them increasingly ineffective for public health uses. This is because the continued presence of biocides in the environment from agricultural spraying encourages the accelerated emergence of resistant strains of mosquitoes and other disease vectors. For example, by 1976, 43 species of anopheline mosquitoes (vectors of malaria) throughout the world had developed resistance to dieldrin, and 24 species were also resistant to DDT. Resistance to these biocides by culicine mosquitoes (vectors of yellow fever, encephalitis, filariasis, and dengue) increased from 19 species in 1968 to 41 species in 1975. Not surprisingly, the incidence of malaria has soared in areas of heavy biocide use. The situation is particularly acute in cotton-growing regions, because cotton production tends to rely so heavily on biocides. Integrated pest management (IPM), the alternative to strictly chemical control, can reduce such problems (Chapter 18).

In 1979, Dr. Noel Brown of UNEP warned of "the specter of a biocide-resistant pest world", with all that such a prognosis implies for food production and the incidence of insect related human diseases. Agricultural planners should respond to this threat by incorporating integrated pest management techniques within agricultural projects as rapidly as feasible. On a more immediate level, the hazard of biocide pollution, poisoning, subclinical intoxication, and accumulation in the tissues of humans, domestic animals, and wildlife can all be reduced through appropriate regulations, education, and health and safety measures. Particularly strict and immediate controls are needed for those biocides known to be highly toxic and/or biologically cumulative.

Counterproductive Effects on Agricultural Production: In tropical soils, many biocides leach or decompose more rapidly than in temperate zone soils. In one study, DDT and organophosphorus biocides either disappeared or were less than 10 percent detectable after three or four years of heavy application. Dieldrin degraded more slowly than DDT in an alkaline sedimentary soil (AVRDC, 1979). Acid soils had somewhat higher residues of most biocides. Higher chemical reaction rates and high levels of biological activity in humid tropical soils have been shown to be responsible for removal of toxic residues (Talekar, et al., 1977). However, many herbicides (which now account for 50 percent by cost or weight of all biocides applied in the world) may persist in soils, undermining the growth of future crops. For example, residues of atrazine, a herbicide used in sorghum cultivation, have been reported to persist in the soil whenever rapid drying of the soil surface in the dry season prevents microbial decomposition.

Improved application methods for biocides can reduce their introduction into the environment. At present, as little as one percent of biocides used actually hits the target organisms (Pimentel and Edwards, 1982). No herbicide can be used which will kill all weeds when the crop is present. In many cases, selective weeding is preferable to using poisons intended to kill all weeds (e.g., arsenicals and sodium chlorate), as certain weeds provide food and cover for the natural predators or parasites of crop pests.

Chemical control of pests often kills economically beneficial insects. For example, chemical destruction of cotton pests also may kill the bees essential to the pollination of adjacent crops. Pollinators are more sensitive to biocides than are many pests. Up to one-third of the world's food supply depends directly or indirectly on insect-pollinated plants (Barker, et al., 1979). No less serious is the use of chemicals which destroy the natural enemies of pests. Furthermore, biocides often kill the predators or parasites of relatively unimportant plant-feeding insects, some of which, when freed from predatory or parasitic pressures, rapidly multiply and become new pests.

Some produce carries unacceptably high biocide residues for human consumption. For example, out of 1,258 samples of produce from Mexico in 1977, 7.2 percent were detained or seized by United States officials for having biocide levels in excess of United States Environmental Protection Agency limits, or for having residues of biocides not permitted for use in the United States (GAO, 1979). Meat contamination resulting from the close proximity of cattle ranches to cotton plantations on the Guatemalan Pacific coast caused estimated export losses to the Guatemalan beef industry of US$1.7 to US$2.0 million in the late 1970s. The economic losses attributable to biocide residues are, however, only one aspect of the biocide problem. Of even greater importance is the public health danger posed by potential contamination of the local food and water in agricultural areas in the tropics.

Direct Poisoning of Humans: WHO reports that 500,000 people are poisoned by biocides each year, of whom at least 5,000 die. The biocide poisoning rate is 13 times higher in developing countries than in the United States (Weir and Schapiro, 1981). Poisonings and sub-clinical intoxication by biocides are common in the prime cotton-growing regions of Central America. ICAITI (1977) reports that in Guatemala, a 10 percent increase in the number of hectares planted with cotton appears to result in a 4.5 percent increase in the biocide poisoning rate. Guatemala had 4,822 reported cases of biocide poisoning with 4 deaths, in a recent one-year period (Davies, et al., 1982). High contamination levels (up to 16 ppm) of DDT in human milk have also been reported by ICAITI in Guatemala. Average DDT levels in cow's milk in Guatemala are 90 times those of levels permitted in the United States. Although the effects on infants are unknown, biocide intake through human milk may initiate the sub-clinical intoxication exhibited by many inhabitants of the Central American cotton-growing region.

Experience in the United States and in Vietnam with TCDD or dioxin (a byproduct of the herbicide 2,4,5-T) shows that even if the principal active ingredient of a biocide may be safe, it is essential to ensure that the same holds true for other components of (or even contaminants in) the formulation. Although legislation can require limitations on such contamination, it can never guarantee safety in use. Occasional large group deaths, numbering several hundred, have been reported after the eating of seed grain intended for planting which had been treated with a mercurial fungicide or other biocide. Contributing factors to biocide poisonings include the close proximity of human dwellings to fields; storage of biocides together with food; use of biocide containers to carry water; lack of clean biocide-free water; insufficiency of protective equipment, along with the casual procedures used by persons directly handling and working in sprayed areas; and the scarcity of information or education (particularly illiteracy) on the part of the rural inhabitants, combined with inadequate regard for employee welfare by their employers. Crop-spraying planes, for example, all too often spray when people are in or near the fields. When biocides are applied by aircraft, typically only 40-80 percent of the load lands on the target crop area, even under ideal application conditions. The remainder drifts off into the atmosphere, contaminating the adjacent environment. About 65 percent of all biocides are applied by aircraft in the United States; in the cotton-growing tropics, the percentage it is probably higher. Education and enforcement of basic health and safety procedures are a crucial first step towards reducing the risks that biocides pose to humans.

Effects on Domestic Animals: Domestic animals are exposed to biocides in high-use regions at least as much as humans. In Egypt, for example, Phosvel (an organophosphate nerve toxin also called leptophos) killed 1,000 water buffalo in 1971. Such deaths of domestic animals can be a severe economic blow to low-income rural families. Farmers may also suffer economic losses through the sub-lethal effects of biocides in depressing meat and egg production. Biocide residues that accumulate in milk can also result in health risks and economic losses.

Effects of Wildlife: The severe effects of biocides on wildlife formed the original basis for restrictions imposed in the United States and other developed countries (Carson, 1962). Throughout the world, biocide-associated reproductive failure continues to threaten many species of birds, such as the Peregrine

Falcon (<u>Falco</u> <u>peregrinus</u>). Exceptionally high levels of DDE, a metabolite of DDT, have been found in freshwater fish in Central America and are likely to exist in estuarine and certain marine species in that region as well. This indicates the tendency of the more persistent biocides to spread in ecosystems that are often quite distant from application sites. Biocides generally accumulate in organisms high up in the food chain; therefore, birds of prey and other carnivores are particularly susceptible.

- 142 -

References

Aguirre Batres, F. and Mazariegoa, J.F. 1977. The growth-environment dilemma. Mazingira, (Special issue)3-4:92-96.

AVRDC, 1979. Progress Report for 1978. Taiwan, Asian Vegetable Research and Development Center, 173 p.

AID, 1977. Environmental impact statement on the AID pest management program. Washington, D.C., AID. 2 Vols. (369 p. and 381 p.)

Anon, 1978. Status of pesticides industry in India. Pesticides (India) 12(4):15-17.

Balasubramaniam, A. 1978. Control of pesticides in Malaysia. Planter (Malaysia) 54(631):600-609.

Barker, R.J., Lehner: Y., and Kunzmann, M.R. 1979. Pesticides and honeybees (Apis mellifera): the danger of microencapsulated formulations. Z. for Naturforschung 34c:153-156.

Brown, A.W.A. 1978. Ecology of pesticides. New York, Wiley, 525 p.

Brown, A.W.A. and Watson, P.L. (eds.) 1977. Pesticide management and insecticide resistence. New York, Academic Press, 638 p.

Carson, R.L. 1962. Silent spring. Boston, Houghton Mifflin, 368p.

Chandler, M.T. 1976. Reducing pesticide hazards to honey bees in tropical East Africa. PANS 22(1):35-42.

Cherrett, J.M. and Sagan, G.R. (eds.) 1976. Origins of pest, parasite, disease and weed problems. Oxford, Blackwell Scientific, 413 p.

Conway, R. A. (ed.) 1982. Environmental risk analysis for chemicals. New York, Van Nostrand, 558 p.

Crosby, D.G. 1973. The fate of pesticides in the environment. Ann. Rev. Plant Physiol. 24:467-492.

Crosby, D.G. 1978. Pesticides in the environment (320-327) in Cleaning our environment, a chemical perspective: report by the committee on environmental improvement. Washington, D.C., American Chemical Society, 457 p.

Davies, J.E., Freed, V.H. and Whittemore, F.W. (eds.) 1982. An agromedical approach to pesticide management: some health and environmental considerations. Miami, Florida, USAID and Univ. Miami School of Medicine:320 p.

Dillon, A.P. (ed.). 1981. Pesticide disposal and detoxification: processes and techniques. Park Ridge, New Jersey, Noyes Publications, 587 p.

Dustman, E.H. and Sticker, L.F. 1969. The occurrence and significance of pesticide residues in wild animals. Ann. N.Y. Acad. Sci.160:162-172.

Edwards, C.A. (ed.) 1973. Environmental pollution by pesticides. New York, Plenum Press, 542 p.

Edwards, C.A. 1977. Environmental aspects of the usage of pesticides in developing countries. Mededelingen van de Faculteit Landbouwwetenschappen, Rijksuniversiteit Gent 42(2):853-868.

Ennis, W.B., Dowler, E.M. and Klassen, W. 1973. Crop protection to increase food supplies. Science 188:593-598.

Environmental Protection Agency, 1974. Strategy of the EPA for controlling the adverse effects of pesticides. Washington, D.C., EPA Office of Pesticide Programs, Office of Hazardous Materials, 36 p.

FAO, 1975. Pesticide requirements in developing countries. Rome, FAO (v.p.).

FAO, 1981. Pesticide residues in food - 1980. Joint FAO/WHO meeting. Rome, FAO Plant Production Protection paper 26:79 p.

Food and Fertilizer Technology Center, 1979. Sensible use of pesticides. Taipei, 250 p.

Fowler, D.L. and Mahar, J.N. 1973. The pesticide review. Washington, D.C., U.S. Department of Agriculture, Agriculture Stabilization and Conservation:58 p.

General Accounting Office, 1979. Report to the Congress of the United States--Better regulation of pesticide exports and pesticide residues in imported food is essential. Washington, D.C. General Accounting Office, 106 p.

Goldstein, J. (ed.) 1978. The least is best pesticide strategy. Emmaus, Pa., Rodale Press, 205 p.

Gunn, D.L. and Stevens, J.G.R. 1976. Pesticides and human welfare. Oxford, Oxford University Press, 278 p.

Hardin, G.J. 1972. Exploring new ethics for survival. New York, Viking, 273 p.

Hill, D.S. 1975. Agricultural insect pests of the tropics and their control. Cambridge (U.K.), Cambridge Univ. Press, 516 p.

ICAITI, 1977. An environmental and economic study of the consequences of pesticide use in Central American cotton production. Guatemala City. ICAITI/UNEP. 273 p.

International Labor Organization, 1977. Safe use of pesticides: Guidelines. Geneva, ILO, 38: 42 p.

International Congress of Pesticides Chemistry, 1979. World food production-environment-pesticides: plenary lectures. London, Pergamon Press, 61 p.

IRRI, 1978. Annual Report for 1977. International Rice Research Institute, Los Banos, Philippines, 548 p.

IRRI, 1979. Annual Report for 1978. International Rice Research Institute, Los Banos, Philippines, 478 p.

IRRI, 1980. Priorities for alleviating soil-related constraints to food production in the tropics. Los Banos, IRRI:468 p.

Khan, M.A.Q. and Bederka, J.P. (eds.) 1974. Survival in toxic environments. New York, Academic Press, 553 p.

Kranz, J., Schmutterer, H. and Koch, W. (eds.) 1977. Diseases, pests and weeds in tropical crops. New York, Wiley, 666 p.

Lal, R., and Verma, S. 1977. Environmental pollution with insecticides used in agriculture in India and measures to combat it. Pesticides (India) 11(6):30-35.

Lawless, E.W., Von Rumker, R. and Ferguson, T.L. 1972. The pollution potential in pesticide manufacturing. Washington, D.C., U.S. Environmental Protection Agency, Office of Water Programs, Pesticide study series 5: 250 p.

Matthews, G.A. 1981. Developments in pesticide application for the small-scale farmer in the tropics. Outlook on Agriculture 10 (7): 345-349.

McEwen, F.L. 1978. Food production--the challenge for pesticides. BioScience 28(12):773-736.

McEwen, F.L. and Stephenson, G.R. 1979. The use and significance of pesticides in the environment. New York, Wiley, 538 p.

Meister Publishing Co. 1976. Farm chemicals handbook. Willoughby, Ohio. (v.p).

Morton, H. (ed.) 1979. The pesticide manual: a world compendium. Croydon (U.K.), British Crop Protection Council, 655 p.

National Academy of Sciences, 1975. Pest control: an assessment of present and alternative technologies. Washington, D.C., NAS: 5 vols.

National Academy of Sciences, 1978. Postharvest food losses in developing countries. Washington, D.C., NAS: 206 p.

National Research Council, 1969. Report of the committee on persistent pesticides. Washington, D.C., USDA Agricultural Research Service, NAS: 34 p.

Organization for Economic Cooperation and Development, 1971. The problems of persistant chemicals: implications of pesticides and other chemicals in the environment. Paris, OECD: 113 p.

Perfect, J. 1980. The environmental impact of DDT in a tropical agro-ecosytem. Ambio 9(1):16-21.

Pimentel, D. 1971. Ecological effects of pesticides on non-target species. Washington, D.C., U.S. Office of Science and Technology. U.S. Government Printing Office, 220 p.

Pimentel, D. 1972. Ecological impact of pesticides. Ithaca, Cornell Univ., Dept. Entomology:27 p.

Pimentel, D. 1973. Extent of pesticide use, food supply and pollution. Journ. N.Y. Entomol. Soc. 81:13-33.

Pimentel, D., Hurd, L.E., Bellotti, A.C., Forster, M.J., Oka, I.N., Sholes, O.D. and Whitman, R.J. 1973. Food production and the energy crisis. Science 182: 443-449.

Pimentel, D. (ed.) 1975. Insects, science and society. New York, Academic Press, 284 p.

Pimentel, D. 1976. World food crisis: energy and pests. Bull. Entomol. Soc. Amer. 22:20-26.

Pimentel, D. 1977. Ecological basis of insect pest, pathogen and weed problems (3-31) in Cherrett, J.M. and Sagan, G.R. (eds.) The Origins of Pest, Parasite, Disease and Weed Problems. Oxford, Blackwell Scientific, 413 p.

Pimentel, D. (ed.) 1978. World food, pest losses, and the environment. Boulder, Colorado, Westview Press, 206 p.

Pimentel, D. and Edwards, C.A. 1982. Pesticides and ecosystems. BioScience, 32(7):595-600.

Pimentel, D. and Perkins, J.H. (eds.) 1980. Pest control: cultural and environmental aspects. Boulder, Colorado, Westview Press, 242 p.

Pimentel, D. and Pimentel, M. 1979. The risks of pesticides. Natural History 88(3):24-33.

Pimentel, D. 1979. Economic benefits and costs of pesticides in crop production. Proc. N.Y.S. Hort. Soc., 124:40-45.

Pimentel, D., et al. 1980. Pesticides: environmental and social costs. (99-158) in Pimentel, D. and Perkins, J.H. (eds.) Pest Control: Cultural and Environmental Aspects. Boulder, Colorado, Westview Press, 242 p.

Riseborough, R. 1978. Patterns of pesticides use in third world countries: implication for conservation policies. Draft paper. Berkeley, California, Bodega Bay Institute, 23 p.

Sheets, J.J. and Pimentel, D. (eds.) 1979. Pesticides: contemporary roles in agriculture, health and the environment. Clifton, N.J., Humana Press, 186 p.

Siddaramappa, R., Tirol, A.C., Seiber, J.N., Heinrichs, E.A., and Watanabe, I., 1978. The degradation of carbofuran in paddy water and flooded soil of untreated and retreated rice fields. J. Environ. Sci. Health, 13(4): 369-380.

Talekar, N.S., Sun, L. T., Lee, E.M. and Chen, J.S. 1977. Persistence of some insecticides in subtropical soils. J. Agri. Food Chem., 25(2): 248-352.

Turtle, E.E. 1979. The assessment of possible environmental impacts of field project activities involving the use of pesticides on industrial crops. Rome, FAO:10 p.

UNDP, 1981. Pesticides on industrial crops: environmental operational guideline 1101. New York, UNDP, G3300-1/TL.1:10 p.

van den Bosch, R. 1978. The pesticide conspiracy. Garden City, New York, Doubleday, 226 p.

von Rumker, R., Conlson, G.A., Lacewell, R.D., Norgaard, R.B. and Parvin, D.W. Horay, F., Casey, J.E., Cooper, J., Grube, A.H. and Ulrich, V. 1975. Evaluation of pest management programs for cotton, peanuts and tobacco in the United States. Washington, D.C., Environmental Protection Agency, 108 p. (Final rept. plus suppl.:623 p.)

Wagner, S.L. 1983. Clinical toxicology of agricultural chemicals. Park Ridge, New Jersey, Noyes Publications, 306 p.

Watson, I.A. 1970. Changes in virulence and population shifts in plant pathogens. Annual Review of Phytopathology 8:209-230.

Way, M.J. 1977. Integrated control--practical realities. Outlook Agri. 9(3):127-135.

Weir, D. and Schapiro, M. 1981. Circle of poison: pesticides and people in a hungry world. San Francisco, Inst. for Food and Development Policy, 101 p.

Wood, B.J. 1971. Development of integrated programmes for pests of tropical perennial crops in Malaysia in Huffaker, C.B. (ed.) Biological Control, New York, Plenum Press, 511 p.

World Health Organization, 1970. Control of pesticides: a survey of existing legislation. Geneva, WHO, 150 p.

World Health Organization, 1973. Safe use of pesticides: Twentieth report of the WHO expert committee on insecticides. Geneva, WHO Technical Report Series, 513:54 p.

World Health Organization, 1975. Chemical and biochemical methodology for the assessment of hazards of pesticides for Man. Report of Scientific Group. Geneva. WHO Technical Report Series, 560:26 p.

World Health Organization, 1976. Pesticide residues in food. Joint FAO/WHO Meeting. Geneva, WHO Technical Report Series 592:45 p.

World Health Organization, 1976. Resistance of vectors and reservoirs of disease to pesticides. Twenty-second Report of the WHO Expert Committee on Insecticides. Geneva, WHO Technical Report Series 585:88 p.

World Bank, 1979. Pesticides -- packaging and labeling: occupational safety and health guidelines. Washington, D.C., World Bank, Office of Environmental Affairs, 5 p.

World Bank, 1979. Pesticides: transportation and distribution: occupational safety and health guidelines. Washington, D.C., World Bank, Office of Environmental Affairs, 3 p.

World Bank, 1979. Pesticides: guidelines for use. Washington, D.C., World Bank, Office of Environmental Affairs, 3 p. plus 2 p. annexes.

World Bank, 1978. Pesticides manufacture: safety and occupational health. Washington, D.C., World Bank, Office of Environmental Affairs, 3 p.

World Bank, 1979. Pesticides - application: occupational safety and health guidelines. Washington, D.C., World Bank, Office of Environmental Affairs, 4 p.

Yates, W.E., Akesson, N.B. and Brazelton, R.W. 1981. Systems for the safe use of pesticides. Outlook on Agriculture, 10, (7).

Youdeowei, A. and Service, M. W. 1983. Pest and vector management in the tropics. New York, Longmans Green, 320 p.

18
Integrated Pest Management

Every chemical pest control program selects for its own
failure. Forty years after the discovery of the first
organochlorine compound (DDT) in 1939, and thirty-five years after
the discovery of the first organophosphorous compounds (TEPP and
parathion) in 1945, about 400 arthropod pest species have become
resistant to the major biocides -- up from 182 resistant pests in
1965. Increase in resistance among these pests has been almost
exponential in recent decades (Bottrell and Smith, 1982). UNEP
(1979) states that selection pressure (i.e., the proportion of the
pest population exposed to selection versus the proportion actually
killed), is the major factor in the development of resistance.
Thus, resistance appears most commonly in economically important
agricultural pests and disease vectors which have long been targets
of chemical control. Where plant disease control has relied heavily
on fungicides, disease organisms are also beginning to show
resistance. In fact, any large-scale chemical control campaign can
be expected eventually to cause the development of resistance in
target organisms. This is exemplified by the increased insect
resistance to dieldrin during 1955-1960, concurrent with the
world-wide WHO program intended to eradicate malaria.

After decades of increasing human and economic costs (e.g.,
Farvar and Milton, 1972) and decreasing benefits brought on by
almost total reliance on simplistic methods of chemical control,
integrated pest management (IPM) has been introduced to remedy these
problems. IPM is recognized by FAO, UNEP, and other governmental,
international, and academic bodies as the most safe, economical, and
sensible way to deal with pest and disease problems. Instead of
relying solely on any single control method, this is a broad
approach capitalizing on the widest possible range of pest mortality
factors. Components of IPM include: determining the thresholds of
economic damage; encouraging or introducing effective parasites and
predators; changing the physio-chemical properties of the plant or
animal to be protected so that pest multiplication rates are
reduced; trapping the pest in artificial devices or in decoy hosts;
enticing the pests into non-productive matings; and many others.
Biocide use remains a component of IPM, but ideally it is only a
last resort when combinations of safer and probably cheaper methods
have failed to confer adequate control.

Because IPM relies on the rational exploitation of a variety of procedures, its use requires careful planning, based on extensive knowledge of the taxonomy, properties, and behavior of the organisms and ecological systems concerned. Indeed, to devise and implement an IPM program of any complexity, scientists, extension workers, and farmers need to cooperate in a concerted effort that provides ample scope for creativity and initiative at all levels. Since the details of implementation are primarily in their control, farmers need to participate actively in several ways: to fine-tune to local conditions the generalized research-based procedures (derived nearly always from experience on a limited number of experimental sites), to ensure a measure of area-wide cooperation where this is vital, and to identify topics needing further research. The proven skill and ingenuity of some peasant farmers in creating complex, sustainable production systems suggests that farmer representatives should be invited to participate in all important phases of the development of new IPM systems.

In a cotton project in Punjab, Haryana, and Maharashtra, India, IPM has made a major impact on production. Farmers have made direct cash savings by basing pest control on need, rather than by predetermined and regular sprayings. Two major cassava pests in Africa are now being controlled by the introduction of their original predators (Chapter 4). The number of success stories in IPM use is growing rapidly.

This chapter outlines the environmental implications of some of the components of IPM, including:

-- maintaining relatively high diversity of agro-ecosystems;

-- use of pest- or disease-resistant (or tolerant) crop species or varieties;

-- manipulation of agronomic practices;

-- encouragement of natural parasites and predators;

-- use of microorganisms;

-- limited use of biocides.

With the exception of chemical biocides, as many components as practicable should be routinely incorporated within the IPM program, so that it need not rely too heavily on any one of them. The use of any single (non-microbial) agent may always lead to the development of some degree of pest tolerance as the result of genetic selection. Progress toward that selection will be slowed and diffused if selection acts on the pest population in the largest number of different directions. Since the possibility of breakdown of any IPM component always exists, it is desirable to search continually for further components which may be integrated as

necessary. Overall, it is clear that IPM requires sustained
research; that agronomic and cultural controls normally require
numerous field trials; and that IPM is intensive in information and
in the use of both skilled and unskilled labor.

Monitoring of Pest Populations: Careful monitoring of
populations of crop pests (along with their natural predators and
parasites) is essential to the success of IPM techniques. To ensure
that such monitoring and the appropriate pest control responses take
place, farmers' meetings, development of IPM methods, and relevant
training should all be well established before the project is
undertaken. Functional monitoring programs should assess the
populations of pests and beneficial parasites and predators, as well
as crop response to attack by pests. The aim should be to replace
the tendency to spray as soon as the first pest appears with a
rational, multi-component approach.

IPM stresses careful planning and the use of available
resources, taking into account both abiotic factors (temperature,
moisture, and light availability) and biotic factors (crops, wild
plants, predators, parasites, and pathogens). "Economic threshold",
the maximum pest population that can be tolerated at a particular
time and place without a resultant economic crop loss, is a key
concept in IPM. No control measure is undertaken in IPM unless the
target pest is present at such a level as to be potentially capable
of causing unacceptable economic losses. When biocide applications
are deemed necessary, the amount of biocide used should be
sufficient to reduce the pest population to a level just below the
economic threshold. Any additional use of biocides beyond this
point unnecessarily accelerates the selection for resistant pest
strains and increases all of the other environmental risks of
biocides (Chapter 17). Factors that influence the economic
threshold level can include the price of biocides, the physiological
stage of crop growth, the crop variety used, and the abundance of
parasites and predators, among others.

The Role of Ecological Diversity in Pest Control: If a
population of plants of a host species is diluted sufficiently by
interspersed non-host plants, the insect or pathogen population
attacking them is less likely to occur at damaging levels. For
example, spruce suffers little from the spruce budworm when enough
of the trees in the stand are taxonomically diverse, such as in a
mixture of hardwoods and other conifers. Similarly, in the wetter
areas of their native Amazonian Brazil, rubber trees in extensive
pure stands probably would be quickly demolished if left to
nature--but when scattered as isolated individuals in natural
rainforest, these trees do not suffer severe insect or disease
loads. In general, if a crop culture can be separated from itself
in both time (by rotation with another crop species or bush fallow)
and space (by interplanting with other crop species), its pest load
is likely to be reduced. This is one of the disadvantages involved
in the planting of monocultures. Similarly, the presence of a dry

season in many tropical areas helps to control many crop pests and
diseases (such as the South American leaf blight, Micocyclus, of
rubber). While it may increase total output, the use of irrigation
can intensify pest or disease problems by effectively eliminating
the natural disruption in pest and disease proliferation that a dry
season can provide (Chapter 20).

Three is the ideal minimum number of rotating vegetation
patterns, if pest populations are to be subjected to selection in
many different directions. However, care is necessary when choosing
the different crop species with which to create this temporal and
spatial diversity. For example, in the Cauca Valley, Colombia,
increasing agricultural diversification with sorghum, maize, and
tomatoes led to increasing bollworm infestation of the valley's
major crop, cotton, because all three crops are hosts of the
bollworm.

Leaving patches of uncultivated (preferably undisturbed)
forest or other natural vegetation among cultivated areas can be a
useful option in IPM. In addition to their aesthetic and
conservation values (Chapters 17 and 24), such patches may provide
refuges for natural enemies of crop pests. However, they may at
times also harbor other crop pests (e.g., insects, wild pigs, birds,
and mistletoes). Thus, while the benefits of uncultivated patches
of natural vegetation are clear from the standpoint of soil and
water conservation, climate amelioration, fuelwood supply, support
for valuable insect pollinators, biological conservation, and
aesthetics, their value in IPM (whether positive or negative) is
difficult to assess without experimentation. The greatest range of
options is produced by aiming for maximum species heterogeneity in
the early phase of agricultural development. As farmers, extension
workers, and researchers learn more of the various species
interactions, unfavorable factors can be minimized by selectively
altering this initial species heterogeneity.

Use of Pest- and Disease-Resistant Crops and Livestock:
Genetic pest- and disease-resistance, one of the major tools of IPM,
is almost the environmentally ideal form of pest control for several
reasons: no biocide residues are left on crops; no biocide
pollution of the environment occurs; minimal harm accrues to
beneficial organisms; and there is no disturbance of the balance
between destructive insects and their natural enemies. Furthermore,
the technique also is compatible with biological, chemical,
cultural, and other control methods. Unfortunately, as with
biocides, the effectiveness of pest- and disease-resistance as a
control usually declines when it is extensively used. Moreover,
raising the levels of protective toxic chemicals in the crops (e.g.,
alkaloids in the potatoes, hydrocyanic acid in cassava) and altering
their nutrient (e.g., vitamin) content can affect their economic
value.

The breakdown of crop resistance is due to processes
similar to the mutation and selection that lead to loss of

effectiveness in biocides. The widespread growing of a resistant variety offers a great advantage to any strain of pest which can overcome the resistance mechanism. Since such vulnerability of a popular crop variety may lead to a sudden demand for biocides, plant breeders should be encouraged to develop varieties protected either by combinations of major resistance genes, or by the less well understood polygenic or "horizontal" resistance (Robinson, 1977).

Durability of resistance is the prime objective. Dependence on varieties protected by only single genes has led to the "boom-and-bust" cycles that have long characterized the breeding of temperate cereals for disease resistance. With the recent efforts to induce pest resistance in rice (Pathak, 1977), a similar pattern of boom-and bust associated with single-gene protection is now recurring. However, whereas a new, single-gene, temperate-zone cereal variety may last five years before breakdown by a rust, a new single-gene rice variety lasted only 18 months in the Solomon Islands.

Where mixed varieties or "multilines", developed through research, meet specific quality and yield objectives, their concurrent use confers more durable protection. However, introduction of multilines without a rational IPM basis may provide ideal conditions for the development of pest or disease strains that are able to overwhelm simultaneously all of the resistance genes present. Where suitable major resistance genes are in short supply, they should be conserved by using all possible additional control methods. Some crop breeding programs have developed multiple sources of pest resistance and maintain alternate genetic solutions to potential pest problems, available for rapid multiplication when necessary.

Alternatively to breeding for physiological incompatibility between crop tissues and pests, durable resistance has sometimes been achieved by reducing the contact between crops and pests, hence deterring pest selection. For example, crop varieties bred to mature early may escape attack; females of a pest species may not be able to lay eggs on a crop variety bearing densely pubescent leaves.

Many traditional, unimproved crop varieties have high levels of the poorly understood polygenic or "horizontal" resistance. Often it is associated, perhaps inextricably, with slow growth rates and modest yield potential. However, sources of polygenic resistance may become critically important if major resistance genes are exhausted. The promotion and spread of ephemerally resistant varieties, however attractive their crop yields may be, causes concern when their use depletes natural reservoirs of potentially more useful polygenic resistance (Chapter 24).

Manipulation of Agronomic Practices: Depending on the nature of the pest, agronomic practices involving sowing date, planting density, fertilizer application, water regulation, crop sequence, and weed control can be used to control pest populations. Such practices act on the pest by affecting the availability and quality of host plants, the presence of cover from predators and parasites, and the impact of unfavorable physical conditions. Thus, a change of planting date alone may advance or retard the crop's period of greatest susceptibility to a pest so that it no longer coincides with climatic conditions favoring a seasonal peak in pest numbers. Cooperative synchronization in the planting of a single crop over a large area can significantly reduce pest damage. This is because if the pest's food supply appears at the same time over a large area, the pest often cannot multiply fast enough to overrun most of the crop before it is harvested or becomes no longer palatable to the pest. Furthermore, if the density of crop plants is reduced, certain pests may suffer higher mortality during dispersal, thus attenuating their population buildup. Planting the intervening spaces with other crop species also reduces pest populations, unless the other crops are also highly attractive to the pests (Perrin, 1977). The use of early-maturing varieties reduces opportunities for pest reproduction, sometimes allowing the crop to escape attack completely. Projects can therefore be environmentally improved by adopting those agronomic practices that will reduce reliance on chemicals. The importance of utilizing research, farmers' past experience, supervision, extension, and training cannot be overemphasized.

Heavy fertilization and abundant irrigation encourage certain types of pest buildup by promoting luxuriant growth of the host plants on which the pests feed. Even a slight reduction of these inputs (and a consequent lowering of the potential yield) may improve pest control. For example, in Peru, a restricted irrigation schedule successfully controls the bollworm, Heliothis virescens, in cotton fields. The natural balance is shifted against the pest by discouraging the succulent growth of leaves and stems which are attractive to egg-laying moths. Furthermore, the useful predator, Paratriphleps laeviusculus, cannot effectively control high pest densities in the humid microclimate of heavily irrigated fields. Since heavy applications of fertilizer and irrigation water can lead to eutrophication and elevated salinity of inland waters, the controlled use of such inputs is also desirable from an environmental standpoint.

Sequencing and rotating botanically unrelated crops has long been an effective procedure to control pests within traditional agricultural systems. Pests which are specific to certain groups of related crops are thus denied the opportunity of continuous increase. However, crop rotation often cannot control pests that are both highly mobile and polyphagous (i.e., eat a fairly wide variety of food). For example, in central Texas, sequentially cultivating alfalfa and maize in the same river valley with cotton encourages the bollworm. The bollworm population feeds on the alfalfa crop in the early spring, increases to great abundance on maize in adjacent fields later in the season, and then, greatly augmented, transfers to cotton during the summer months. Sometimes a regionally agreed-upon, crop-free time gap can control pest outbreaks and diminish the heavy biocide dependence that frequently accompanies continuous, high-input cropping.

While many weed species are agronomically significant only because they compete with crop plants for resources, other species have been identified as harboring pests and diseases. Lists of weed species potentially responsible for carry-over of nematodes, insect pests, and virus diseases and their insect vectors are often available. If crop monitoring suggests that these pests or diseases could cause economically significant losses, special attention should be given to controlling the relevant weeds. Maize production in the United States is the country's largest user of herbicides and the second largest user of insecticides (after cotton). These biocides are used primarily against the maize rootworm complex that can be effectively controlled by rotating maize with legumes or small grains. In 1945, before biocides were commonly used, most maize in the United States was grown under such rotations.

Pest Control with Parasites and Predators: There are essentially two ways of exploiting natural enemies to control pest populations. The first is through a management system that preserves and augments naturally existing predators and parasites. The second is through mass-rearing of natural enemies for release, either as nuclei for further reproduction in the field or in sufficiently large numbers to regulate a pest population directly (without the need for further natural breeding). Numerous examples exist demonstrating the effectiveness of natural predators and parasites. For example, the egg parasite Trichogramma, abundant in cotton-producing areas of Southern California, is capable of destroying 45-50 percent of cabbage looper and bollworm eggs during one growing season. Various genera of spiders (Tetragnatha, Clubiona, Araneus, Oxyopes, and Tylorida) are natural predators that can effectively suppress stemborers, adult gall midges, leafhoppers, and planthoppers in rice cultivation.

In IPM, several measures help to preserve and augment natural enemies:

(1) In general, biocides should be restricted for use only in the absence of suitable alternatives, as there are no species-specific biocides.

(2) Biocides, if employed, should be chosen not only for their effects on the pests concerned, but also for their ability to spare as many beneficial insects or other organisms as possible.

(3) Habitats for predator species should be created or preserved, either by retention of sufficient natural vegetation or by using crop species that serve as suitable hosts.

(4) Predators and parasites can be encouraged either by regulating moisture levels to provide suitable habitat (through appropriate irrigation regulation), by leaving some weeds which meet their needs for shelter, and by supplying food (even the pest itself) when natural prey or hosts are absent or scarce.

- 155 -

(5) Natural enemies can be promoted and protected. For example, the People's Republic of China recently banned the catching, selling, and buying of frogs and toads in an effort to promote them as insect control agents. Three years of experimentation demonstrated that by encouraging frogs, the production brigades could cut the amount and cost of chemical biocides by more than half. In Malaysia, Barn Owls (Tyto alba) are being encouraged as rat control agents in oil palm plantings.

Rearing beneficial organisms for mass release can be done on whole plants or parts, or on acceptable simulated (fictitious) hosts, or on artificial media. Thus augmented, populations of ladybird beetles, for example, may successfully control aphids.

When the pest system is well enough understood, it may be advantageous to introduce pest populations deliberately at critical times of the year in order to maintain sufficient numbers of predators or parasites. In any case, effective monitoring of populations of key species is usually essential for successful IPM.

Microbial Control of Pests: Naturally occurring insect pathogens (e.g., bacteria, fungi, protozoans, viruses, and nematodes) can help control many pests. Besides causing outright death, insect pathogens also may interfere with insect development and reproduction, or reduce their resistance to attack by parasites, predators, or other pathogens. Pathogens may also influence the susceptibility of insects to control by chemical biocides. In some cases, chemical control of plant pathogens is counterproductive because it also destroys insect pathogens.

FAO (1973) stated that microbial control has extremely high potential because micro-organisms:

(1) are highly selective;

(2) are generally safer to handle than chemical biocides;

(3) leave no residues or pollution;

(4) are often compatible with chemical biocides;

(5) can often be prepared easily and cheaply; and

(6) can be applied by a variety of methods.

However, certain viruses, bacteria, protozoa, fungi or nematodes have the potential to harm people, livestock, or natural enemies. Fortunately, those microbial preparations presently in commercial use are considered generally safe for humans and most other animals, and research into new "microbial insecticides" also seeks to ensure that they will be safe to use, even in large-scale operations.

In the long run, microbial control techniques may be of utmost importance because they are less subject than most other pest control techniques to eventual diminution of their effectiveness by the development of resistance in crop pests. Most pest control techniques, including chemical biocides, use of crop varieties specially bred for pest resistance or tolerance, use of sex pheromones or other lures, and possibly even the mass-release of sterilized male pests, are vulnerable in varying degrees to a common flaw--their tendency to promote the emergence of resistant pest strains. However, microbial pathogens of crop pests continually co-evolve in nature with their hosts. Thus, the microbes maintain their ability to control pest outbreaks whenever the microbes are taken from the field, mass-reared, and released in large numbers. Natural predators also co-evolve with crop pests, but may be less able to "keep pace" with the pests.

Chemical Control as Part of the IPM System: Any use of synthetic biocides in IPM should be based on a weighing of the possible pest-control benefits against the numerous potential adverse environmental effects. These include biocide residues on crops, contamination of waterways, health hazards to humans, and destruction of wildlife, including pollinators and other beneficial insects (Chapter 17).

Among the very few narrowly-selective biocides that have been developed so far are picrimicarb for aphids and dimethirimol for cucurbit powdery mildew. But even these have resulted in the selection of resistant strains; all other chemical biocides can be expected ultimately to produce similar results. For the time being, chemical biocides can be used to best effect by avoiding use except when absolutely necessary, and then only in a carefully modulated range of dosages, formulations, and dates and methods of application. In general, non-persistent, naturally derived substances, such as pyrethrins (recently supplemented by synthetic analogs), are less environmentally damaging than the traditional synthetic biocides.

Integrated pest management does not aim for pest eradication (which may not be desirable and is almost never possible anyway). In IPM, biocides are used only when economic damage is likely. This practice has the added benefit of tilting the balance in favor of the pests' natural enemies. Furthermore, the development of resistance to biocides is likely to be slowed if less rigorous selection occurs as a result of less intense spraying. It follows that the most lethal biocides often are not the most effective ones in the longer run.

Other Agents Useful in IPM: There are many plants which have insecticidal properties. Water extracts from the Indian Neem Tree, quassia, derris, pyrethrums, and the tobacco plant have all been used effectively. However, either because of limited availability, very short persistence, high toxicity, or high cost, many of these have been superseded by modern synthetic substitutes. However, research into plant species with insecticidal properties continues (Rosenthal and Janzen, 1979). The rapidly disappearing tropical rainforests contain many thousands of plant species, some of which are known to contain chemicals capable of controlling insect populations; habitat preservation and accelerated research are urgently required in these areas. Ethnobiological studies of the strategies developed by indigenous peoples to control insect pests with natural biota can provide exceptionally valuable information for modern applications in pest management.

Light oils derived from petroleum continue to be effective for the control of several insect and mite species. Such oils are only minimally polluting. The National Institute of Science and Technology in the Philippines recently reported success in producing a low-cost, pollution free biocide from coconut oil. The new biocide, when applied on the leaves of tomatoes, onions, and leafy vegetables, was found to be effective against corn borers, aphids, and moth larvae. Coconut oil degrades easily and does not harm most other living species (Development Forum, 1979). In a parallel finding from Egypt, a powerful, water-soluble molluscicide, of major potential in schistosomiasis control, can be prepared as an infusion of the ragweed **Ambrosia maritima** (IDRC, 1978).

References

Adam, A.V. 1976. Pesticide development and need in developing countries. Washington, D.C., Proc. XV Int. Congr. Entomol. 741-746.

Adam, A.V. 1977. Importance of pesticide application equipment and related field practices in developing countries (217-225) in Watson, D.L. and Brown, A.W.A. (eds.) Pesticide Management and Insecticide Resistance, London, Academic Press, 638 p.

Allen, G.E. and Bath, J.E. 1980. The conceptual and institutional aspects of integrated pest management. BioScience 30(10): 658-664.

Allen, G.E. et al. (eds.) 1978. Microbial control of insect pests: future strategies in pest management systems. Gainesville, Florida, NSF--USDA--Univ. Florida, 290 p.

Apple, J. L. and Smith, R.F. 1972. A preliminary study of crop protection problems in selected Latin American countries. Washington, D.C., USAID, 41 p.

Apple, J. L. and Smith, R.F. (eds.) 1976. Integrated pest management. New York, Plenum Press, 200 p.

Baker, K.F. and Cook, R.J. 1974. Biological control of plant pathogens. San Francisco, W.H. Freeman, 433 p.

Barfield, C.S. and Stimac, J.L. 1980. Pest management: an entomological perspective. BioScience 30(10):683-689.

Batra, S.W.T. 1982. Biological control in agroecosystems. Science 215:134-139.

Beingola, O. 1974. Integrated pest control Latin America. (77-94) in Environment and Development. Indianapolis, SCOPE Misc. Publ. 412 p.

Birch, M. (ed.) 1974. Pheromones. Amsterdam, North Holland Publishing Co., 495 p.

Bottrell, D. G. 1979. Integrated pest management. Washington D.C., Council on Environmental Quality: 041-04-0049-1.

Bottrell, D.G. and Smith, R.F. 1982. Integrated pest management. Environmental Science Technol. 16(S):282A-288A.

Boza, B. 1972. Ecological consequences of pesticide used for the control of cotton insects in Canete Valley, Peru. (423-38) in Milton, J.P. and Farvar, T. (eds.) The Careless Technology. New York, Garden City, Natl. Hist. Press, 1030 p.

Brader, L. 1979. Integrated pest control in the developing world. Ann. Rev. Entomol. 24:225-254.

Clausen, C.P. (ed.) 1978. Introduced parasites and predators of arthropod pests and weeds: a world report. Washington, D.C., USDA Agricultural Handbook 480:545 p.

Council on Environmental Quality, 1980. Integrated pest management (Integragency Coordinating Committee). Washington, D.C., US Government Printing Office, 41 p.

DeBach, P. 1974. Biological control by natural enemies. London, Cambridge University Press, 323 p.

DeBach, P. (ed.) 1974. Biological control of insect pests and weeds. New York, Halsted Press, 825 p.

Development Forum, 1979. New Filipino pesticide. Rome, FAO, Division of Economics and Social Information 7(7):4.

Durham, W.F. and Williams, G.H. 1972. Mutagenic, teratogenic and carcinogenic properties of pesticides. Ann. Rev. Entomol. 17:123-148.

Falcon, L.A. and Smith, R.F. 1973. Guidelines for integrated control of cotton insect pests. Rome, FAO:92 p.

FAO, 1980. Collaborative action in strengthening plant protection. Rome, FAO:14 p.

FAO, 1978. Integrated pest control: Report of the eighth session of the FAO panel of experts held in Rome (4-8 Sept): 32 p.

Farvar, T.M. and Milton, J.P. (eds.) 1972. The careless technology. New York, Natural History Press, 1030 p.

Flint, M.L. and Bosch, R. van den. 1981. Introduction to integrated pest management. New York, Plenum Press, 256 p.

Glass, E.H. (ed.) 1975. Integrated pest management:rationale, potential, needs, and implementation. Entomol. Soc. of Amer. Spec. Publ. 75-2:141 p.

Huffaker, C.B. (ed.) 1980. New technology of pest control. New York, Wiley, 530 p.

Huffaker, C.B. and Messenger, P.S. (eds.) 1976. The theory and practice of biological control. New York, Academic Press, 788 p.

IOBC, 1981. Future trends of integrated pest management. International Organization for Biological Control of Noxious Animals and Plants (IOBC)(Bellagio 1980 Conference) London, Center for Overseas Pest Research.

IDRC, 1978. Searching. Ottawa, International Development Research Council, 32 p.

Jacobson, M. 1971. Naturally occurring insecticides. New York, Marcel Dekker, 585 p.

Jacobson, M. 1972. Insect sex pheromones. New York, Academic Press, 382 p.

Jacobson, M. (ed.) 1975. Insecticides of the future. New York, Marcel Dekker, 93 p.

Laird, M. and Miles, J.W. 1983. Integrated mosquito control methodologies. London, Academic, 355 p.

Maxwell, F.G. et al. 1972. Resistance of plants to insects. Adv. Agron. 24:187-265.

Metcalf, R.L. and Luckmann W. (eds.) 1975. Introduction to insect pest management. New York, Wiley, 587 p.

Metcalf, R.L. 1980. Changing role of insecticides in crop protection. Ann. Rev. Entomol. 25:219-256.

National Academy of Sciences, 1975. Pest control: an assessment of present and alternative technologies. Washington, D.C., NAS. 5 vols.

National Academy of Sciences, 1974. Productive agriculture and a quality environment. Washington, D.C., Committee on Agriculture and the Environment, NAS, 189 p.

National Academy of Sciences, 1977. Insect control in the People's Republic of China. Washington, D.C., Committee on Scholarly Communication with the People's Republic of China, NAS, 218 p.

Office of Technology Assessment, 1979. Pest management strategies and crop protection. Washington, D.C., US Govt. Print. Off. 2 vols.

Pathak, M.D. 1977. Defence of the rice crop against market pests. Annals N.Y. Acad. Sci. 287:287-295.

Perkins, J.H. 1982. Insects, experts and the insecticide crisis: the quest for new pest management strategies. New York, Plenum, 324 p.

Perrin, R.M. 1977. Pest management in multicropping systems. Agro-ecosystems 3:93-118.

Pimentel, D. and Goodman, N. 1979. Ecological basis for the management of insect populations. Oikos 30(3):422-437.

Pimentel, D. 1980. Environmental risks associated with biological controls (11-24) in Lundholm, B. et al. (eds.). Environmental protection and biological forms of control of pest organisms. Stockholm, Ecol. Bull.31.

Pimentel, D. (ed.) 1981. Handbook of pest management in agriculture. Boca Raton, Fla. CRC Press Handbook Series, 3 vols.

Prayoon, D. et al., 1974. Integrated control of cotton pests in Thailand. Bangkok, Thailand, Ministry of Agriculture. Plant Protection Service, technical bulletin 32:27 p.

Rajamohan, N. and Jayaraj, S. 1978. Field efficacy of Bacillus thuringiensis and other insecticides against pests of cabbage. Indian J. of Agricultural Sciences 48(11):672-675.

Ridgeway, R.L. and Vinson, S.B. (eds.) 1977. Biological control by augmenting natural enemies. New York, Plenum, 416 p.

Robinson, R.A. 1977. Plant pathosystems. Annals NY Acad. Sci. 287:238-242.

Rosenthal, G. A. and Janzen, D. H. 1979. Herbivores: their interaction with secondary plant metabolites. New York, Academic, 718 p.

Smith, E.H. and Pimentel, D. (eds.) 1978. Pest control strategies. New York, Academic Press, 334 p.

Staples, R.C. and Toenniessen, G.H. 1981. Plant disease control: resistance and susceptibility. New York, Wiley, 312 p.

Southwood, T.R.E. and Way, M.J. 1970. Ecological background to pest management. (6-29) in Rabb, R.L. and Guthrie, F.E. (eds.) Concepts of Pest Management. Raleigh, North Carolina State University Press, 242 p.

UNEP, 1979. The state of the world environment: the 1979 report of the executive director. New York, United Nations. 14 p.

van den Bosch, R., Messenger, P.S. and Gutierrez, A.P. 1982. An introduction to biological control. New York, Plenum, 262 p.

van Rumker, R. 1974. Production, distribution, use and environmental impact potential of selected pesticides. Washington, D.C., U.S.D.A. Office of Pesticide Programs, 439 p.

Winteringham, F.W. 1975. Fate and significance of chemical pesticides: an appraisal in the context of integrated control. EPPO Bulletin 5:65-71.

Wood, B.J. 1971. Development of integrated control for pests for tropical perennial crops in Malaysia. (422-457) in Huffaker, C.B. (ed.) Biological Control. New York, Plenum, 530 p.

19
Tsetse Control
in Livestock Projects

This chapter outlines the environmental disadvantages associated with tsetse fly (Glossina) "eradication" 1/ components in livestock projects. If, after factoring these considerations into overall project design, tsetse control in some sites is deemed unavoidable, then some environmentally less harmful control methods are suggested. In any tsetse control projects that employ toxic chemicals, especially persistent biocides, a biological and chemical monitoring component should be used to reduce the risks of induced pest resistance and non-target damage and deaths. Furthermore, significant research support for environmentally preferable alternatives to chemical control is important to benefit the design and reduce the risks of future projects.

Background--Range Management: Much semi-arid land being considered for increased livestock range cannot sustain even present livestock levels. Since much, if not most, tsetse-occupied land is of poor quality, most tsetse control is for extensive livestock ranching, rather than for agriculture. Animal husbandry recognizes that increased production is not synonymous with increased stocking. In fact, the true wealth of extensive livestock ranching lies in the range, rather than in the animals themselves. Nonetheless, compelling political and economic reasons may be adduced for dense stocking: communal grazing discourages herd control and pasture improvement, nomads have little use for consumer goods which are not easily transportable, and persons lacking formal

---

1/ The term tsetse "eradication" in the sense of "to eliminate completely" is inappropriate since eradication is probably impossible and unlikely to be fully desirable. Tsetse control in certain regions, to certain levels compatible with whatever other activities are contemplated and for a certain finite duration may be attempted, with a more reasonable likelihood of achievement.

education traditionally rely upon their herds, not holdings of currency or accounts in modern banks, to provide security during hard times. Dependence on milk can also create ecological imbalances between the pastoralists and their stock on the one hand, and the range environment on the other. Excessive numbers of stock (i.e., above the carrying capacity of the range) are often maintained as an insurance against dry years and because production is somewhat risky (Jewell, 1980). By selling animals earlier and finishing them elsewhere (on improved pastures), production can be increased, but at lower stocking levels. Range deterioration can be halted and recovery promoted by proportionately reducing grazing pressure and implementing seasonal rotation and other management programs. While necessarily stringent management techniques are more likely to be adopted by large, "modern" cattle companies, such high-technology projects are less likely to benefit directly the rural poor of a region.

A major goal of development is to promote sustainable and maximally efficient land use, so that people may prosper within constraints of the resources available in their ecosystems. Continuing abuse of grazing lands over large parts of Africa has exacerbated the effects of droughts and contributed to famine, malnutrition, and related problems. As is known all too well, overgrazing intensified the recent Sahelian disasters. Supervision and training components in development projects help to deter overgrazing (especially in areas with significant potential for desertification). Special care is needed in any livestock projects within Africa's pivotal 10^0 N belt. The environmentally preferred approach is to develop land management compatible with prevailing environmental conditions, rather than the usual converse attempt to modify the environment to suit domestic cattle. Savanna and semi-arid regions can be made continuously productive if accorded more judicious management. Preparation and implementation of a rational land-use plan related to a project area's carrying capacity can effectively address most environmental concerns. Use of full-time range management specialists can prevent land degradation caused by overgrazing.

Game Ranching as an Environmentally Preferable Alternative to Tsetse Eradication:

Proponents claim that game ranching is superior to cattle ranching because it is cheaper, requires less (or no) tsetse control, and lowers major capital expenditures. Mixed-species game herds (including domesticated Lechwe and Eland) optimize available forage, since such herds consume a correspondingly larger spectrum of vegetation (both graze and browse) from more plant species, from different heights, at different times of year, and at different intensities. Even mixed-species domesticated livestock (e.g., cattle, sheep, and goats) are typically less damaging than monospecific herds. Indigenous ungulates require smaller ranges, consume less water, reproduce faster, and mature sooner than do introduced domestic stock. Cattle exert heavier pressure on the environment because they have to trek to water nearly every day, and they graze more than they browse. Furthermore, cattle ranching may displace indigenous fauna. Jewell (1980) and Gruvel (1980) outline ways to make livestock and wildlife more compatible. With appropriate management, meat production (by weight) from native fauna can be significant, as pointed out by such specialists as David Hopcraft (1980, 1981), Hugh Lamprey (1974, 1979), the Mossmans (1976), and Lambrecht (1972). Hopcraft (1981) summarized well the promise that game ranching offers:

"....utilization of wildlife for meat and hide production is the only alternative we have to our present systems of domestic stock ranching. And more importantly, it is a land-use system that by its nature is nondestructive. It...is the key to the survival of those seriously threatened degrading lands... Ninety-five percent of all arid and semi-arid grasslands of Africa are threatened (by the syndrome of monoculture cattle/overstocking/erosion/desertification).... In land-use three considerations should be primary. First, the landuse system chosen should be as close as possible to the natural system... Second, we should always move towards multiculture (e.g. 10-20 species of ungulates). Third, we must use systems that require no imported energy because these are expensive and, in the long term, impractical. Cattle ranching meets none of the above criteria. Game ranching, on the other hand, fits perfectly..."

With game ranching, tsetse-control expenditures are vastly reduced, being restricted to limited areas, such as villages and river fords. To be successful in a locality, game ranching should be specifically designed for the local environment. Thorough testing and modification on a pilot scale is prudent before a large investment is made in a full-scale operation.

Other Land Uses Compatible With Tsetse Flies: In certain cases, it may be desirable to promote non-grazing forms of land use, such as cultivation of tree or shrub perennial crops, to the extent they do not promote tsetse. (However, tsetse has, on occasion, adapted to pine plantations in Kenya.) Especially promising areas in this respect are found within Cameroon, Chad, Ethiopia, Guinea, Ivory Coast, Mali, Niger, Nigeria, Sudan, Upper Volta, and the Central African Republic.

Trypano-tolerant Cattle: If no environmentally preferable alternatives to cattle appear feasible, then a pilot project using trypano-tolerant cattle may be tested. Lee and Maurice (1983) consider use of trypano tolerance to be the most economical option for tsetse areas. Trypano tolerant cattle such as the N'Dama and Baole breeds also resist such fly-belt diseases as streptothricosis and tick-borne fever. Under medium levels of management on range with a carrying capacity of 2-4 ha per animal unit, these cattle can be expected to achieve a 70 percent calf drop with only 5-8 percent mortality. Such breeds produce 500 lb grass-fattened steers in 2-2.5 years and achieve a 15 percent offtake rate, equalling zebu cattle. These figures are still improving although, whether they can be maintained through normal dry season cycles has yet to be determined. A productivity index recently developed by the International Livestock Center for Africa (ILCA) and based on reproduction, viability of cows and calves, milk production, growth rate, and live weight of adult cows, indicates that trypano-tolerant cattle breeds can be as productive as non-trypano-tolerant ones.

Conservation of trypano-tolerant cattle breeds threatened with extinction is needed to preserves humanity's options for the future. Crossbreeding with exotics, improved utilization of the various breeds and their crossbreeds in humid and semi-humid tropical regions, and research on trypano-tolerant livestock and trypano tolerance generally are also important. Research is currently being done to investigate the origins of trypano tolerance, and to determine whether it is immunological, genetic, or both. As with all cattle herding, degradation of the pastoral resource by overgrazing, erosion, and depletion of land fertility is a major risk in the absence of proper precautions.

Trypanocides and Vaccines: Chemical prophylaxis and therapy of tolerant and susceptible cattle is already useful, but it is impaired by trypanosomal resistance to all major drugs. Vaccines for humans and cattle are being actively developed. Using game ranching while awaiting availability of an appropriate cattle vaccine may turn out to be a wise course. The International Laboratory for Research on Animal Diseases (ILRAD) in Nairobi provides the latest treatment methods and products; it also publicizes information concerning efforts to develop effective and reliable experimental vaccines. In the longer term, trypanosomiasis control appears to be more practical and environmentally less damaging than tsetse eradication campaigns.

Tsetse Control Without Biocides: Integrated pest management (IPM, Chapter 18) is the combined application of biological, mechanical, chemical, and other measures to control pests, carried out in the context of adequate ecological understanding of the particular problem and region. One IPM component, discriminative bush clearing, has sometimes been effective in tsetse control in the past and promises to become increasingly effective as tsetse ecology becomes better known. Removal of all bush in 1-2 kilometer wide "barriers" reduces tsetse crossings. Fences can reduce the problem of game (which may carry flies) crossing such barriers. To annihilate wildlife on a large scale is expensive, brutal, and generally unacceptable among civilized people today; it is also wasteful of a valuable natural resource and of questionable effectiveness in tsetse control (Regan, 1982). Release of specially-bred sterile male tsetse flies already has proven a useful procedure in Tanga, Tanzania; this method is likely to become even more effective as research progresses. Disorientation by odor attractants (e.g., carbon dioxide and acetone), pheromones, and sound emissions also has proven useful and environmentally benign in pilot schemes. Mechanical fly traps have been effective on a local scale (e.g., in the 150 km^2 island of Princile in the Gulf of Guinea); improved designs are likely to improve their effectiveness (Lee and Maurice, 1983).

The Glossina predator candidates for tsetse control include hersiliid spiders and asiliid flies, which can decimate adult tsetse populations, and some ants and carnivorous Coleoptera which attack tsetse pupae. Tsetse pupae are also killed by a variety of parasitoids. The most effective of these are Mutilla wasps, which can infect over 50 percent of tsetse pupae in an area. Since Mutilla is common in eastern, central, and southern Africa, it should be tried in West Africa without delay. Phoretic mites and nematode worms also are promising control candidates, as are some fungal and microbial pathogens. These or other specific possibilities for non-chemical control should be included within first-stage trials of tsetse control projects.

Historical hindsight indicates that the retention of forest might have prevented the spread of certain Glossina species and their trypanosome parasites. In Sierra Leone, the replacement of forest with savanna, farms, bush fallow, and oil palm plantations over the past century and a half has caused changes in the distribution and frequency of several Glossina species and the diseases they carry. Since many Glossina species thrive in more open conditions, they tend to spread with extensive forest clearing (Payne, 1980).

Tsetse Control Using Biocides: Dieldrin and endosulfan (i.e., thiodan) are both broad-spectrum chlorinated hydrocarbons similar to DDT (Chapter 17). In practice, these biocides also kill non-target organisms, such as birds (including non-insectivorous species), mammals, reptiles, amphibians, and fish. They also damage even phytoplankton and soil bacteria, as well as decimating beneficial insects. Furthermore, their residues tend to persist in the soil for 6 to 8 (or even as much as 15) years. Populations of insectivorous birds, hitherto efficient predators of tsetse flies, are severely depleted and in some cases obliterated as a result of dieldrin or endosulfan applications. Since dieldrin has been completely proscribed in most temperate countries (e.g., in the United States by the Environmental Protection Agency), promoting the use of such a dangerous poison in the tropics could be construed as ethically questionable. Since endosulfan is relatively more toxic to tsetse flies than dieldrin, a smaller quantity needs to be used, with commensurately fewer resulting risks. The tolerance of Glossina to biocides is a potential problem in any project relying on these chemicals (Chapters 17 and 18).

Figure 19.1: Environmental Ranking of Biocide Choices

Rank	Biocide	Remarks
1.	Pyrethroids	Least risk; least persistent; narrower spectrum of organisms affected.
2.	Endosulfan	Risky; less persistent; broad-spectrum but more toxic to Glossina.
3.	Dieldrin	Most risk; persistent; broad-spectrum.

Note: If project designers become convinced that tsetse control and cattle ranching projects are the most prudent forms of investment, then less damaging options for tsetse control should be substituted for the often proposed dieldrin.

Methods of Biocide Application: The most hazardous method of biocide application is aerial dispersal, whether from fixed-wing aircraft or helicopters. Environmental risks are reduced somewhat by the "knock-down" method, in which low dosages of endosulfan are applied in ultra-low volume (ULV) aerial applications. Nontheless, aerial spraying, however modified, remains more dangerous than discriminatory ground spraying. Only 50-80 percent of aerial spray usually reaches the target area, the remainder drifting elsewhere in the environment. Although ULV uses a concentrated spray of extremely small droplets to improve coverage, this technique increases the hazard of drift. Recent development of the "charged droplet" techniques offers potential for increasing the proportion of the spray which lands on the target. In any case, the quantity of biocide used is much less when applied from the ground than from the air. Aerial spraying is also much more energy-intensive, making it even less desirable for developing countries that must expend scarce foreign exchange for petroleum imports (Chapter 21).

Figure 19.2: Environmental Ranking of Biocide Application Methods

1. Least damage: Discriminative and selective ground spraying (backpacks).

2. More damage: Mistblower from trucks.

3. Still more damage: Helicopter dispersal.

4. Most damage: Fixed-wing aircraft dispersal.

Monitoring of Biocide Effects: Should the other technical factors related to project design force the use of persistent biocides, then a component for both chemical and biological monitoring should be an essential part of the project, in order to help mitigate resulting problems. Monitoring of biocide residues in water, air, soil, bottom sediments of water bodies, and plant and animal tissues can provide a useful early warning system. Biological monitoring should also be used to detect changes in plant and animal populations or physical well-being before, during, and after the project. The abundance of key indicator species of insectivorous birds (e.g., *Muscicapidae* flycatchers) is worth emphasizing in such monitoring. For maximum effectiveness, both chemical and biological monitoring should begin at least one, and preferably several, breeding seasons before biocide application begins.

References

de Vos, A. 1978. Must Africa suffer the environmental consequences of tsetse-fly control? Unasylva 30 (121):18-24.

FAO, 1978. Game as food. Unasylva 29:1-36.

FAO, 1978. Report of the second consultation on the programme for the control of African animal trypanosomiasis. Rome, FAO: 75 p.

FAO, 1980. Trypano-tolerant livestock in West and Central Africa. Rome, FAO Animal Prod. Health Paper 20:2 vols.

Goldsmith, E. 1976. Editorial letter to the Director of FAO. The Ecologist 6 (6): 198-199.

Gruvel, J. 1980. Glossina, domestic livestock and wild fauna: is it possible to reconcile them? Acta Zoologica et Pathologica Antverpiensis 75:29-48.

Hopcraft, D. 1970. East Africa: the advantage of farming game. Span 13(1):29-32.

Hopcraft, D. 1980. The natural land-use system of wildlife ranching. New York, Vital Speeches (May):485-489.

Hopcraft, D. 1981a. Wildlife ranching in perspective (68-71) in Karstad, L. et al. (eds.) Wildlife Disease Research and Economic Development. Ottawa, IDRC, 179e:80 p.

Hopcraft, D. 1981b. Nature's technology (211-224) in Coomer, J. (ed.). Quest for a sustainable society. New York, Pergamon, 260 p.

IDRC, 1974. Tsetse control: the role of pathogens, parasites. and predators. Ottawa, IDRC, No. 034C:22 p.

Jenkins, D.W. 1964. Pathogens, parasites and predators of medically important arthropods. Geneva, World Health Bulletin Supplement 30:150 p.

Jewell, P.A. 1980. Ecology and management of game animals and domestic livestock in African savannas (353-381) in Harris, D.R. (ed.) Human Ecology in Savanna Environments. London, Academic, 522 p.

Jordan, R.M. 1979. Trypanosomiasis control and land use in Africa. Outlook on Agriculture, 10(3):122-129.

Koeman, J.H., Balk, F., and Takken, W. 1980. The environmental impact of tsetse control operations. Rome, FAO, Animal Production Health Paper 7:71 p.

Laird, M. 1977. The future for biological methods in integrated (Tsetse) control. Ottawa, Int. Dev. Research Cent. 220 p.

Lambrecht, F.L. 1972. The Tsetse fly: A blessing or a curse? (726-741) in Farvar, T.M. and Milton, J.P. (eds.) The Careless Technology. New York, Natural History Press, 1030 p.

Lamprey, H.F. 1974. Management of flora and fauna in national parks (235-358) in Elliot, H. (ed.) Second World Conference on National Parks. Morges, IUCN, 504 p.

Lamprey, H.F. 1979. Structure and functioning of the semi-arid grazing land ecosystem of the Serengeti region, Tanzania (562-601) in Tropical Grazing Land Ecosystems. Paris, UNESCO, 522 p.

Lee, C.W. and Maurice, J.M. (eds.) 1983. The African trypanosomiasis: methods and concepts of control and eradication in relation to development. London, Centre for Overseas Pest Research/Washington, D.C., World Bank, Tech. Paper 4: 107 p.

McKelvey, J.J. 1973. Man against tsetse. Ithaca, New York, Cornell University, 306 p.

Mossman, S.L. and Mossman, A.S. 1976. Wildlife utilization and game ranching. Morges, International Union for the Conservation of Nature and Natural Resources, Occasional Paper 17.

Ormerod, W.E. 1976. Ecological effect of control of African Trypanosomiasis. Science 191:815-821.

Regan, T. 1982. All that dwell therein. Berkeley, California University Press, 249 p.

Payne, I. 1980. Trees and disease. New Scientist (January 3):12-15.

Shafer, M. 1977. Tsetse fly eradication and its implications. Washington, D.C., USAID Technical Assistance Bureau, No. 147-590:54 p.

Taylor, C.R. 1969. The eland and oryx. Scientific American 220:88-95.

USDA, 1978. Trypanosomiasis bibliography. Hyattsville, Md. Emergency Programs Veterinary Services, 182 p.

20
Irrigation

The expansion of irrigation and increased efficiency in the use of irrigation water are among the most important opportunities for improving tropical agricultural productivity. In many tropical regions, only a small fraction of cropped land is irrigated. For example, irrigation accounts for only 3 and 13 percent of the cropped land in Africa and Latin America, respectively. Although lack of rainfall limits agriculture on 44 percent of the land in Africa, and on 32 and 17 percent of the land in Central and South America, respectively, many such areas lack available water bodies which can be tapped economically for irrigation.

Irrigation--perhaps the largest single subsector of agricultural development projects--can have many complex and long-term environmental impacts. These include the initial and immediate impact of the heavy machinery used to clear and grade land, with the resulting loss of habitat for flora and fauna (Chapter 24). The contruction of dams, canals, and ditches and the impoundment and diversion of water bodies involve additional problems, including the loss of land. The long-term, adverse effects of irrigation schemes also may include soil waterlogging, salinity, and alkalinity, unless adequate provisions are made for drainage and leaching. Increased incidence of waterborne diseases, water pollution, and crop pest and disease occurrence all can result from large, irrigated oligocultures (involving one or a few specialized crops). Such negative impacts can be prevented or reduced by careful environmental planning and feasibility studies, and through subsequent environmental monitoring.

Diversion of water resources for irrigation projects can involve major impacts. The following sections outline the different environmental effects of water diversions directly from rivers and streams, storage behind dams in surface reservoirs, and pumping of groundwater.

Direct Water Diversion: The same streams and rivers that are diverted for irrigation are also frequently used as conduits for wastes. In dry seasons, a river's base flow after diversion may cease to be sufficient to support downstream or instream (e.g., fisheries) uses, or to dilute wastes to acceptable levels. In the

United States, for example, where only about 10 percent of all
cropland is irrigated, about 35 percent of the water pumped from
rivers is used for irrigation; less than half of this amount returns
to rivers. Besides the obviously harmful effects on downstream
users, such flow reduction can create breeding places for pests and
diseases in the stagnant pools that remain. Dry season monitoring
of waste disposal in rivers and of the quantity of water taken for
irrigation can mitigate some problems, although the enforcement of
water quality regulations is frequently difficult. Run-off and
drainage canals from fields sometimes contain non-biodegradable
agro-chemicals, which may be concentrated by evaporation or
bioaccumulation to dangerous levels. Fertilizers in irrigation
run-off water can cause eutrophication of downstream water bodies.
Simple water diversion often adversely affects fish, wildlife, and
downstream vegetation.

Storage in Surface Reservoirs: Surface water storage
causes environmental changes in the area of impoundment as well as
downstream; some of these changes are damaging. The losses of
forests, interred minerals, and fertile land due to reservoir
inundation are environmental and economic costs that need to be
weighed against the benefits generated by irrigation. However,
extraction of valuable timber species prior to impoundment reduces
economic losses. Because it reduces the quantity of biomass which
will decay while submerged, such extraction often improves water
quality. Other adverse effects of reservoirs include loss of water
by evaporation, raising of adjacent water tables, and possible
salinization. The inundation of forests or other natural habitats
frequently implies the destruction of unique ecosystems and
consequent extinction of plant and animal species (Chapters 21 and
24; Goodland, 1979). Reservoirs can be designed and managed to deny
breeding habitats to most potential disease vectors, such as insects
and snails, but the economic costs may prove prohibitive. Vector
surveillance and control teams are necessary in all projects that
increase disease vector breeding opportunities.

Proliferation of water weeds in reservoirs reduces the
oxygen content of water, sometimes to the point of killing fish.
Water weeds also impede navigation and fishing from boats, increase
water loss through evapo-transpiration, and provide disease vector
breeding sites. Chemical control of water weeds is likely to
provide only temporary relief (Chapter 23).

Silt entrapment behind large dams reduces downstream
sediment loads. Although decreased sediment loads in irrigation
canals reduce operation and maintenance costs, reduced sediment
loads in rivers can imply increased scouring tendency, hastening the
erosion of river banks and structures (e.g., bridge foundations),
and, in some cases, of deltas. Moreover, decreased nutrient loads
have caused declines in riverine, estuarine, and marine fisheries.
Reservoir water in new impoundments can become depleted of oxygen
because of decomposition of flooded biomass (e.g., forest). This
can cause fish kills downstream when deoxygenated water is released

through the dam. Dams also create barriers to the migration of both freshwater and marine (anadromous) fish species, thereby further reducing fishery potential.

Pumping of Groundwater: Overpumping is the major environmental concern associated with the use of groundwater for irrigation. The safe level of pumping is determined by the availability of water for recharge, the transmissibility of the aquifer, and the potential for aquifer contamination by surface pollutants, upwelling of connate brines, or seawater intrusion. Where groundwater is pumped to provide water for livestock, determination of the carrying capacity of the surrounding range and control of stocking rates helps to avoid overgrazing and, in certain regions, desertification.

Saltwater seepage into coastal aquifers and the resulting aquifer collapse are almost impossible to reverse. Careful monitoring and control of pumping rates, as well as region-wide regulation of the new well drilling, is often needed to prevent excessive groundwater withdrawal and the resulting deterioration of the aquifer. In some areas, the presence of mangrove swamps markedly reduces saltwater intrusion; the protection of these unique ecosystems should be an important priority.

Energy Costs of Irrigation: A very large energy input is needed to deliver the large volumes of water needed for large-scale irrigation schemes. About 10 million liters (or some 10,000 tons of water by weight) are needed to irrigate each hectare of semi-arid land. In fact, the energy needed to pump water for irrigation may double or even quadruple the total energy inputs required for producing a crop (Chapter 20).

Water Conservation: Although surface irrigation requires less energy than either sprinkle or trickle irrigation systems in applying the same quantity of water, it usually uses significantly more water. For some crops, water consumption using sprinkle and trickle irrigation systems may be 30 to 60 percent less than with surface irrigation. "Dead-level" surface irrigation provides water savings comparable to sprinkle and trickle irrigation systems, but it also may require a higher energy input than for normal surface irrigation systems.

Trickle irrigation not only conserves water, but can also minimize such problems as salinization and waterlogging. However, the added immediate energy and installation costs of the delivery system must be balanced against the long-term value of water conservation. Furthermore, some water must still go through the root zone annually to leach away the accumulated salts.

Waterlogging and Salinization Due to Over-irrigation and Poor Drainage: Waterlogging in irrigated cropland is caused by excessive water application and/or poor drainage. Such saturation immediately harms crops, causing stunted plant growth, delayed crop maturity, and shallow root development. Longer-term agro-environmental problems, such as raised water tables and soil salinization and alkalinization, can be extremely severe. Prior to the construction of irrigation systems, land suitability studies should focus upon soil permeability and drainage potential. In this manner, such problems can often be routinely avoided.

Beyond routinely providing drains and other mechanical means for removing excess water and for leaching salts, well-run irrigation schemes regulate the quantity and even the quality of the irrigation water used (e.g., as recommended in the United States Federal Water Quality Guidelines). Because irrigation water has often traditionally cost much less than other agricultural inputs, farmers tend to over-irrigate when the water supply permits; this often results in waterlogging, waste of water, and deterioration of soil. Procedures for efficiently pricing irrigation water are an important, although politically sensitive, option to increase water use efficiency and to reduce environmental costs.

Sedimentation and Erosion: Sediment deposition on fields in small amounts can be beneficial to crops. However, sedimentation of irrigation works and excessive silt deposit on fields can cause extensive damage. In this case, mismanagement of the upstream environment harms irrigation systems; irrigation systems themselves normally do not contribute to siltation. Upstream soil conservation measures, particularly conservation of existing forest and reforestation of critical catchment areas, are the most effective, long-term means to control sedimentation. Careful management of the upstream watershed is important to providing a relatively steady water supply, as well as to protect the irrigation investment from sedimentation.

Erosion resulting from poor irrigation system design can seriously reduce the depth of fertile topsoil. Fast-running irrigation water, runoff from heavy rainfall on nearby slopes, or improperly terraced ground stripped of vegetation can all lead to the erosion of not only topsoil, but also fertilizer, seedlings, and young crop plants. Properly designed terraces and furrows, along with other soil conservation measures, will reduce erosion by limiting the rate and quantity of water runoff.

Soil and Water Pollution: The intensification of agricultural production through irrigation is closely linked with increased and sometimes excessive use of biocides and fertilizers. Agro-chemical run-off and seepage can pollute surface and ground water. Irrigation with waste water, environmentally a recommended practice, must be well managed to protect workers and grazing animals from disease and to prevent the heavy metals present in industrial and urban waste water from accumulating in soils and contaminating crops (FAO, 1977).

Biocides: Some biocides may reach groundwater tables in irrigated areas. The rate of breakdown of biocides in groundwater is extremely slow because there are no microbes to aid decomposition. Biocides that run off into surface waters can harm aquatic fauna and make many species unfit for human consumption. During project design and implementation, efforts should be made to minimize the environmental costs of reduced water quality and threats to aquatic life resulting from biocide runoff.

Careless use of biocides in irrigation schemes can induce biocide resistance among pests. Intensive monocultures or oligocultures over large contiguous areas in irrigation schemes can boost plant pest and disease populations and create many breeding places for insect and other vectors of human and animal diseases. When malaria or other insect-borne diseases break out, biocides used in excessive amounts to control crop pests often become ineffective against the newly-resistant populations of disease vectors. As a result, disease incidence in irrigated areas can soar to epidemic proportions (Chapter 16; Giglioli, 1979).

Fertilizers: Application of fertilizers increases nitrate (NO_3) levels in water. In agricultural regions of Israel, for example, NO_3 concentrations in some wells have reached 100 mg/l. The permissible limit is about 40 mg/l. Although chemical fertilizers are still used fairly sparingly in tropical countries, excessive water flows may wash fertilizer residues out of irrigated fields and into waterways. In such cases, eutrophication of waterways is likely to occur, along with its attendant consequences -- increased aquatic plant growth, clogged waterways, disruption of aquatic fauna, and increased breeding sites for insects. While fertilizer run-off should be reduced for economic as well as environmental reasons, eutrophic waterways can be kept at least partially clear with such biological control measures as the introduction of herbivorous fishes. Some fishes can also effectively control insect larval populations in irrigation waterways (Chapter 15).

Irrigation with Waste Water: Irrigation with waste water is environmentally beneficial because water resources are conserved, organic matter and nutrients are recycled, and pollution is mitigated. Although environmental health hazards can ensue if waste management is inadequate, recycling of waste water is strongly recommended where feasible. Pathogens (including fungi, bacteria, protozoa, viruses, and helminths), and heavy metals and other chemicals from waste water can survive in soils and harm irrigated crops. Proper treatment of waste water, however, minimizes this risk. No disease outbreaks have yet been linked to properly treated wastes. Farmworkers exposed to <u>untreated</u> waste water are particularly vulnerable to parasitic infections; regular medical attention and use of adequate footwear on farms irrigated with waste water lessen the possibilities for such infections. Although sunlight and air-drying during spray irrigation with waste water help to kill bacteria and viruses, spraying also creates aerosols which may spread disease organisms downwind over considerable distances. This problem can be avoided by establishing a buffer

zone between farms using waste water and nearby residential areas.
When done with inadequate precautions, sewage irrigation of pastures
has led to increased tubercular and tapeworm infections in cattle.
The World Health Organization has promulgated a set of regulations
and standards for waste water irrigation in order to prevent such
infections (WHO, 1973, 1979, 1980).

**Salinity due to Irrigation Run-off and Drainage
Effluent**: When water is repeatedly diverted from a river for
irrigation purposes and eventually returned, salinity levels
inevitably rise. As a classic example, the Colorado River at the
United States-Mexico border is rapidly approaching a brackish state
(United Nations, 1976). The increased--and increasing--salinity of
the water (already 1,500 mg/l) is primarily due to the leaching of
natural minerals from the soil, inherent in irrigation.
Desalinization, an extremely costly solution, consumes much energy
and is not completely effective in desalting the water.
Nonetheless, to honor its agreement with Mexico to provide
low-salinity water in the Colorado River, the United States is
constructing the largest desalting plant in the world at Yuma,
Arizona, with an initial construction cost of approximately US$356
million. The expense is justified by the calculation that each
milligram per liter of salt in excess of 115 mg/l would cause damage
to downstream agriculture worth US$230,000, and the projection that,
by the year 2000, the salt content of the river at the Mexican
border would be 1,300 mg/l. A large quantity of highly concentrated
saline liquid remains to be disposed of after the desalinization
process. With proper precautions, saline effluent can often
feasibly be diverted to sumps for solar evaporation or directly
discharged into the ocean.

Pests and Weeds in Irrigation Projects: The very large
but often unappreciated economic value of a dry fallow season (or a
winter) is the suspension of active competition by insect pests,
plant pathogens, and weed growth in cropped fields. Insect
populations drop radically, microbial pathogens cease growth, and
most weeds revert to seeds because of a lack of water. To the
extent that irrigation projects provide water during the dry season,
pests and weeds can be expected to continue proliferating.

Once an irrigation system is implemented, the pests can be
controlled by turning off the water at certain key periods during
the dry season. Instead of irrigating throughout the dry season, a
relatively brief period of no irrigation can be maintained long
enough to reduce pest pressure to acceptable levels (Chapter 18).
Sometimes, irrigation periods can be manipulated to maintain pest
populations at levels that are too low to cause economic damage, but
high enough to provide a year-round food supply for the natural
predators or parasites which keep their population in check.
Alternatively, these natural enemies may be re-introduced during
each new growing season. Such pest management options depend on the
drought resistance of the crop, the water needs of natural pest
predators, the crop rotation used, the crop ripening time, and the
tradeoffs between foregone crop growth and foregone biocide use.

Interruption of irrigation has been demonstrably successful in controlling mosquito and snail infestations. Occasionally the reverse applies: where sufficient water is available, insects and weeds can also be controlled by judicious flooding. A good example of this is flooding to control weeds in paddy rice.

Irrigation and Human Diseases: Without strict control measures, the expansion of irrigation can lead to increased incidence of water-borne diseases such as schistosomiasis and mosquito-borne diseases such as malaria, encephalitis, and yellow fever. Other water-related diseases, including diarrheal and gastrointestinal disorders, cholera, and typhoid, are also closely linked to irrigation and other water development schemes.

Schistosomiasis vector snails breed in the slow-flowing water common in irrigation canals and drains. Engineering measures can be taken to reduce snail populations by increasing flow rates. Particular irrigation methods, such as sprinkler irrigation, can discourage spread of the disease to a certain extent, but the problem of snails in open canal distribution systems remains .

Onchocerciasis (river blindness) is primarily associated with the fast-flowing, oxygenated water found in sluices, weirs, and spillways. It affected entire villages in Upper Volta after the construction of major rice irrigation schemes (Finkel, 1979). Since the black fly that spreads onchocerciasis depends upon sunlit areas of fast-flowing water for breeding, keeping relevant irrigation works under roofing or other shade will minimize this hazard.

The incidence of malaria, encephalitis, and yellow fever is related to the availability of wet areas suitable for mosquito breeding and the the presence of reliable food sources for mosquitoes, such as domestic animals and humans on irrigated farms. Among the most attractive mosquito breeding sites are stagnant ponds, undrained syphons and culverts which remain unflushed during irrigation off-seasons, and marshy areas resulting from seepage and overtopping of canals. All of these conditions can be corrected by good engineering. Influxes of infected migrant workers into areas where **Anopheles** mosquitoes are present, but where there is no history of malaria, can trigger the rapid spread of malaria as the vectors pick up and transmit the disease-causing plasmodium parasite. Any use of biocides, regardless of the agricultural justification, undermines their public health effectiveness, as mosquitoes readily acquire resistance. This problem can be reduced by area-wide coordination to use separate types of biocides for health and agriculture. Predatory fish, oil suffocation, and other means can be used to control mosquito larvae in waterways, thereby reducing reliance on biocides.

References

Arceivala, S.J. 1977. Water re-use in India in: Schuval, H.I., (ed.), Water renovation and re-use. New York, Academic Press, 303 p.

ATDO, n.d. Under-soil irrigation with pitcher and PVC pipe: alternative techniques for water scarcity areas. Appropriate Technology Development Organization, Government of Pakistan, Islamabad.

Berg, R. van den, Hagan, R.M., and Kovda, V.A. (eds.), 1973. Irrigation, drainage and salinity: international source book. FAO/UNESCO. London, Hutchinson, 510 p.

FAO, 1977. Organic materials and soil productivity. Rome, FAO Soils Bulletin 35: 119 p.

Feachem, R., McGarry, M., and Mara, D. (eds.) 1977. Water, wastes and health in hot climates. New York, Wiley, 399 p.

Finkel, H.J. 1979. Guidelines: The environmental impacts of irrigation in arid and semi-arid regions. Rome, FAO: 41 p.

Giglioli, M.E.C. 1979. Irrigation, Anophelism and Malaria in Adana, Turkey. Washington, World Bank, 86 p.

Goodland, R. 1979. Environmental optimization in hydrodevelopment of tropical forest regions (10-20) in Panday, R.S. (ed.). Man-made Lakes and Human Health. Parimaribo, Univ. Surinam, 73 p.

Hagan, R.M., Haise, H.R., and Edminster, T.W., (eds.) 1967. Irrigation of agricultural lands. Madison, Wisconsin. American Society of Agronomy, 1,180 p.

Hotes, F.L. and Pearson, E.C., 1976. Effects of irrigation on water quality. Symposium in Arid Lands Irrigation in Developing Countries, Environmental Problems and Effects. Alexandria, Egypt, 44 p.

Kirkpatrick, F.W. and Smythe, E.F. 1967. History and a possible future of multiplier re-use of sewage effluent at the Odessa, Texas, Industrial Complex. in: Cecil, L.K., (ed.), Water Reuse (Chemical Engineering Progress Symposium Series) New York, American Institute of Chemical Engineers 63:201 p.

Maas, A. 1978. .... And the desert shall rejoice; conflict, growth, and justice in arid environments. Cambridge, Mass., MIT Press, 447 p.

McJunkin, F.E. 1975. Water, engineers, development, and disease in the tropics. Washington, D.C., Agency for International Development, 182 p.

- 180 -

National Academy of Sciences, 1974. More water for arid lands:
promising technologies and research opportunities. Washington,
D.C., NAS:154 p.

Peters, W.B. 1979. The role of water and land resource information
in world Bank programs for agricultural development. Workshop
on Soil Resources Inventories and Development Soil Resource
Study Group, Agronomy Department, Cornell University; Ithaca,
New York, 38 p.

Smith, M.A. 1979. Re-use of water in Australia. in: Water
Resource Symposium Proceedings. Denver, CO, AWWA Research
Foundation, Vol. 2: 925-936.

Tillman, R.E. 1981a. Environmental guidelines for irrigation.
Washington, D.C., U.S. Agency for International Development:
74 p.

Tillman, R.E. 1981b. Environmentally sound small-scale water
projects. New York, Codel Inc., 142 p.

United Nations, 1976. Increasing available water supplies through
weather modification and desalination. United Nations, New
York, Water Resources Branch, (E/CONF. 70 TP 191): 14 p.

Wasbotten, T.P. 1978. Public health and nuisance considerations
for sludge and wastewater application to agricultural land.
in: Knezek, B.D. and Miller, R.H. (eds.) Application of sludges
and wastewaters on agricultural lands: a planning and education
guide. Washington, D.C., US Environmental Protection Agency,
91 p.

WHO, 1973. Reuse of effluents: methods of wastewater treatment and
health safeguards. Geneva, WHO Technical Report Series, No.
517:33 p.

WHO, 1979. Human viruses in water, wastewaster and soil. Geneva,
WHO Technical Report Series No. 639: 50 p.

WHO, 1980. Health aspects of treated sewage re-use. Geneva,
(Algiers) WHO: 43 p.

21
Energy in Agriculture

Now that energy is no longer cheap, it ranks in importance with the classical factors of production--land, labor and capital--and its more limited availability is of concern to agriculture. More efficient use of commercial energy is essential for tropical countries wanting to approach any measure of self-sufficiency in food production. Energy helps boost yields per unit of land and enables human labor to be more productive, thus aiding the twin foundations of economic improvement in agriculture. Both direct and indirect uses of fossil fuels in agriculture are increasing (Green, 1978). However, since commercial energy is likely to become even more expensive in the future as oil supplies are depleted, agricultural systems should avoid excessive dependence upon commercial energy-based inputs. The transition from living off energy _capital_ (fossil fuels) on borrowed time to living off energy _income_ (mostly solar-based) is inescapable; conscious planning to bring it about is preferable to the severe economic dislocations and human suffering that an unplanned transition would surely bring.

As an example of extreme energy intensiveness, twelve or more tons of oil equivalent are needed to land a single edible ton of shrimp in Mexico and elsewhere. The chain of enterprises involved in feeding people in Europe and North America now requires about 400 gallons of oil equivalent per person per year. This is three times the average per capita use of commercial fuels for all purposes in the developing world (Leach, 1976). While fossil fuel inputs of this order may still be economically tolerable in developed countries, they cannot continue indefinitely. Moreover, they are an inappropriate model for most tropical countries, which already face severe balance of payments problems stemming greatly from their dependence on imported petroleum.

This chapter addresses the energy dilemma of tropical agriculture. Although energy use still must increase to help meet the most basic human needs of still expanding populations, the tropics possess little spare capital for energy-intensive, high-technology solutions to their agricultural problems. Rather than remaining in a state of "fuel-poverty", or buying supplies of dwindling fossil fuels at crippling prices, developing countries will be more secure to the extent they rely even more heavily on indigenous fuel sources, such as mini-hydropower, wood, the fermentation products of vegetative and waste materials (e.g., alcohol and biogas), and solar energy. Through diversification of energy sources, recycling, and general natural resource conservation, major improvement in the current energy situation is possible. Following the energy-intensive path of modern agriculture, however, can only lead to deep impoverishment in most developing countries. A judicious mix of high labor input, low petroleum input, and heavier reliance on solar-based technologies may be the most prudent approach.

Table 21.1: Farm Energy Inputs in Modern Industrial Agriculture

Item	Percentage Average	Percentage Range
Mechanization (including fuel)	50	3-70
Fertilizers	30	20-70
Irrigation	5	0-50
Biocides	10	0-40
Miscellaneous	5	2-40
	100	

Note: Percentage of total energy used for specific activities per hectare of crop production.
Source: Pimentel (1979;1980a, b).

Table 21.1 presents, in approximate descending order, the largest sources of commercial energy use in agriculture. Since the ranking varies by country, region, and prevailing cropping patterns, only general and qualitative observations and options are discussed below.

Mechanization: About half of the energy used to raise crops in industrialized countries goes into machinery and the fuel needed for its operation. Such mechanization has greatly reduced the necessary labor input to produce a given harvest. Mechanization does not necessarily increase yields per unit of land, but can allow farmers to optimize the time spent on cultivation and harvest and to deal better with some marginal soil conditions (e.g., those that require pan-busting).

The advantage provided by extremely large tractors (175 horsepower) doing more work per unit time than smaller (15-50 horsepower) models is offset by their much greater fuel consumption during operation in existing-sized fields. The vast fields of monocultures owned by agri-business are suitable for large tractors much more than the more diverse holdings of smallholder farmers. A tradeoff exists between expending more fuel in large tractors or more time in smaller, more energy-efficient tractors. Because labor is usually relatively much cheaper than commercial energy in the tropics, small farmers should, as a rule, be encouraged to use smaller machines. Furthermore, on small holdings, where field layout requires more turning, the better maneuverability of small tractors is advantageous.

Although mechanization, if wisely employed, may contribute to accelerated seed bed preparation, the removal of indurated horizons, and the provision of supplemental moisture in some situations, machines are environmentally detrimental when they compact the soil, reduce water availability, and increase susceptibility of the soil to wind and water erosion. Such damage can be reduced through proper selection of machinery and by using skilled machinery operators. Hand-held tillers, walking mini-tractors, and small tractors are more easily serviced at the village level. When one breaks down, the loss is less severe (and shorter) than if a large tractor breaks down.

Some light tasks can be carried out with small amounts of labor, rather than machines; this approach can significantly reduce energy inputs. For example, for one person to apply biocides to a hectare of land by hand sprayer requires less than 1,000 kilocalories, whereas the same task carried out with that person driving a 50-horsepower tractor requires more than fifty times this amount of energy. The former method is quite practicable for small areas of annual crops, or for larger areas of perennial tree crops when the time schedule is not so pressing.

Although this book does not directly address most of the social issues related to tropical agriculture (see Introduction), one such issue is mentioned here because of its often serious environmental repercussions. One of the negative consequences of agricultural mechanization in countries with large amounts of "surplus labor" is an increase in unemployment (or underemployment). Such mechanization often occurs despite relatively cheaper labor for a number of institutional and socio-political reasons, including subsidized credit, the "prestige value" of modern machinery, and landlords' fear of unionization or other collective action by agricultural laborers. When low-income peasants are displaced by mechanization and fail to find suitable alternative employment, they frequently seek to colonize new, agriculturally marginal lands (e.g., steep hillsides and forested or semi-arid areas). Severe environmental disruption often results, including deforestation, soil erosion, increased flooding and landslides, desertification, and loss of habitat for rare species, among other problems. Moreover, the severe poverty these people face remain unalleviated, in part because the newly-occupied marginal lands often cannot sustain more than 2-3 meager crop harvests before they must be abandoned (Ledec, 1983). Policies that serve to alleviate poverty, promote land reform, increase rural employment opportunities, and discourage excessively labor-displacing mechanization deserve support because of their environmental, as well as their social, benefits.

Crop Innovations: One promising area for energy conservation is the cultivation of food-bearing tree crops as a partial alternative to annual crops (SERI, 1980). The breadfruit tree, _Artocarpus incisa_, for example, is a traditional food crop, but it is not fully exploited, even though one tree can yield one ton of fruit per annum for many years with no energy-intensive inputs and minimal labor requirements. This is more than the yield from two "average" cereal crops, which are not feasible in many environments and which also require far more labor and other inputs. The pupunha or pejibaye palm of Latin America, Guilielma gasipaes, is another tree that provides a staple food that is rich in oil or starch (depending on the genotype used). These and other species are being developed for intensified cultivation in both West Africa and Latin America. Their main advantage is that once planted and bearing (at three to five years from planting), they continue to produce food for decades without any further cultivation at all. Thus, they offer major energy savings when compared with annual crops and are environmentally better suited to most tropical environments, since they protect the soil from erosion and sustain a relatively closed nutrient cycle (akin to that of a native forest).

Fertilizers: The worldwide return on the use of additional fertilizer is diminishing (Brown, 1983). Nonetheless, for most crops, the single largest input of commercial energy is for inorganic "chemical" fertilizers (mainly nitrogen, phosphorus, and potassium). Nitrogen fertilizer in particular is the most energy-intensive input, requiring nearly half a gallon (1.6 liters) of gasoline equivalent to produce, process, and package 1 kilogram. In maize cultivation, for example, nitrogen fertilizer is applied at a rate of about 130 kg/ha (accounting for about 190 liters of gasoline equivalent).

FAO's 1979 target rate of agricultural production growth of 3.5 to 4.0 percent per year implies an expansion in fertilizer use of 9 percent per annum. The implication that the world can or should double the number of fertilizer factories within eight years is extremely serious, given the likely continued increase in commercial energy prices. Moreover, fertilizers, particularly nitrogenous ones, also pollute the stratosphere. To the extent nitrogen compounds from fertilizer production deplete the world's protective ozone shield, a painful tradeoff may have to be faced before long.

Since nitrogen fertilizer is water soluble, severe leaching losses can be reduced by using synchronized and split application methods. This not only facilitates the fullest use of the fertilizer for crop production but also reduces the environmental pollution caused by nitrogen runoff. Nitrogen fertilizer use can also be reduced when nitrogen-fixing legumes are planted in association with the main crops. For example, a common and successful combination in Latin America is the cropping of beans (_Phaseolus vulgaris_) with maize; the legume enhances soil nitrogen, while also helping to control corn-root worm and reduce weed competition. In irrigated rice, alternation of the rice with a dry season soybean crop helps to conserve soil nitrogen and is also beneficial to soil texture. While recognizing the benefit of such combinations, it is important to note that legumes will also compete for soil nutrients, including nitrogen, until the symbiotic assimilation of soil nitrogen in their root nodules develops sufficiently to keep the plant supplied with nitrogen.

The potential for utilizing biological fixation of nitrogen is particularly high in rice, with the water fern Azolla showing great promise as a partial substitute for nitrogen fertilizer. Atmospheric nitrogen is assimilated by the nitrogen-fixing blue-green alga Anabaena azollae, which associates symbiotically with Azolla. Studies by the International Rice Research Institute (IRRI) show nitrogen fixation rates of 120 kg/ha in 106 days. The fern may be grown before the rice crop and incorporated into the soil at transplanting; alternatively, it may be introduced into the paddy after transplanting. Azolla is widely distributed in rice-growing areas of the tropics and temperate zones and has been proven to promote significant increases in rice yields in California and North Vietnam. Although growth of the fern is depressed at temperatures higher than 31°C, high temperature-tolerant Azolla strains are being developed.

For tropical tree crops, the creeping legumes Pueraria phaseoloides, Centrosema pubescens, Calopogonium mucunoides, Calopogonium coeruleum, and Desmodium ovalifolium are useful as ground covers under newly established plantings (e.g., of rubber and oil palm). They not only protect the soil from sun and rain, but also fix large amounts of nitrogen that are eventually returned to the soil in leaf, stem, and root residues. These legumes decay quickly, allowing mineral nitrogen uptake by the crop, with a consequent reduction in the need for fertilizer application.

Irrigation: Irrigation, as outlined in Chapter 20, can be an energy intensive system for crop production. As the price of commercial energy continues to rise relative to the value of crops, some crops will no longer be profitably produced under irrigation. This has already happened in some regions where it is no longer profitable to produce alfalfa under irrigation because alfalfa is a relatively low-value crop.

Increasingly, energy-intensive irrigation will be limited to high value crops and for relief of occasional drought in otherwise well-watered regions. Indeed, one or two applications of water during a dry spell in a usually rain-fed area can significantly increase yields or even save the crop during a critical one- or two-week period, should this be economically worthwhile.

Biocides: The production of biocides requires massive energy inputs; it takes about 9 liters of oil equivalent to produce, formulate, package, and transport one kilogram of biocide. For some crops, such as cotton, vegetables, and fruits, large quantities of biocides are used. In addition, as mentioned in Chapter 17, biocides can be a hazard to public health and the environment. Hence, the use of integrated pest management (IPM) technology (Chapter 18) will both reduce energy use and improve environmental quality. The natural parasites and predators used in IPM are truly solar-powered, since they obtain all their energy directly from the crop pest population.

Weeds can be controlled effectively either by hand or mechanical removal or tilling, by biocide application, or by a combination of these. As for energy expenditure, biocidal weed control requires about the same amount of energy as plowing or other forms of mechanical weed control. Intercropping, such as growing mungbeans between corn rows, is also an effective weed control practice, conserving biocides and labor, providing a more diverse and reliable yield, and increasing soil fertility.

As noted above, biocides are an environmental threat. Hand and mechanical tilling can also pose environmental threats, especially when exposing the soil to erosion. Therefore, the environmental and energy factors can be, to some extent, tradeoffs of the various weed control options.

Crop Drying: In intensive agriculture, it is common to harvest grain directly in the field before the grain is fully dry. Then, to keep the grain from spoiling, large inputs of energy are used to dry the crop. Traditionally, maize (as an example) has been harvested on the cob and placed in a corn crib to dry; this system of harvest requires only wind energy to dry the corn, rather than expensive fossil fuel energy. Crop drying systems that use wind, direct exposure to the sun, or solar-powered crop driers are environmentally preferable to those that require commercial energy inputs.

Crop Residues: Proposals have been advanced for the use of crop residues in biomass energy conversion to alcohol and biogas. The Gobar gas fertilizer system, widespread in India, uses mainly manure with some crop residues. Except on flat land (0-2 percent slope), the removal of crop residues can intensify erosion and nutrient depletion, especially in hot, high-rainfall areas.

Crop residues left in place help preserve soil structure and organic matter, deter soil erosion, improve the soil's water-holding capacity, and help prevent nutrient loss due to runoff. In addition, crop residues provide valuable nutrients. For example, a hectare of maize stover contains about 60 kg of nitrogen, 6 kg of phosphorus, and 60 kg of potassium. A large energy input (equivalent to about 100 liters of petroleum) would be required to replace these nutrients with artificial fertilizers. Thus, crop residues left in place are valuable in maintaining the quality of the agricultural environment and in conserving energy.

Livestock Manure: Notwithstanding the fact that most livestock manure is applied to the land, much manure remains inefficiently utilized. For example, in Bangladesh, only 50 percent of the dung from the country's 26 million cattle, if converted to biogas, would supply 75 percent of the country's current energy demand for cooking. The residual sludge from biogas production would provide roughly 50 percent of current fertilizer consumption.

Worldwide, an enormous amount of manure is burned annually for cooking and heating. The approximately 70 million tons of dung burned annually as fuel in India have a soil nutrient content equivalent to one third of India's chemical fertilizer use (FAO, 1970). It is conservatively estimated that worldwide, about 50 percent of the nutrient content of manure (particularly nitrogen) is lost when the manure is spread directly from the barn or stall onto the land. During the fallow season, sunlight oxidizes the nitrogen and rain leaches it from the manure and washes it into streams and lakes. This process constitutes a substantial loss of nitrogen and other nutrients from the soil; moreover, the leached nutrients can seriously pollute water.

More effective use of livestock manure offers major potential gains. Small, gravity-fed ponds or holding tanks lined with impermeable clay (or possibly plastic) can store manure and urine during the non-growing season for application during the growing season. Construction of suitably-sized holding tanks and the addition of another task (spreading the manure early in the growing season when the farmer is busy cultivating and planting) are the major costs of this technology. However, as fossil energy costs rise, the benefits of energy conservation make the extra effort increasingly worthwhile.

The cost of commercial energy for storage of manure and urine in a holding tank is relatively small when compared with the total amount of such energy that is expended in crop production. The major energy costs of using manure involve hauling and spreading. Within a radius of 0.8 to 1.6 km, about 4.6 liters of fuel are expended per ton of manure for hauling and spreading. Although to haul and spread the manure from 2.5 dairy cows to one hectare of land requires about 1 million kcal, about 2 million kcal are saved in foregoing the fertilizer that would be required to replace the soil nutrients present in the manure. Thus, a net saving of 1 million kcal per hectare is possible using livestock manure instead of commercial fertilizers, whenever the manure is relatively nearby. A major limitation of this system, however, is finding an alternative energy source for the manure-supplied heat used for cooking (Chapter 12).

Tradeoff of Land and Energy: Within limits, land and energy resources can be substituted for one another. For example, one hectare can be fertilized sufficiently to produce the same quantity of maize as two hectares would normally yield. The commercial energy input needed to double maize production in this manner is about three times as much as would be required without the extra fertilization.

Choice of Food Crops and Livestock Feeding Systems: Still another option is to grow plants and animals that require the least investment in commercial energy. Although this may not be practical or acceptable in all regions or cultures, much can be learned by examining differential energy requirements among various crops and livestock. Based on a comparison of energy input and the food calorie energy content of the crop, grains are generally more energy efficient than either vegetables or fruits. As mentioned above, two to three calories of food are produced per calorie invested in grain production (because of the extra solar energy contribution). With fruits and vegetables, however, about two calories of energy must be invested by humans to produce one calorie of food energy. These comparisons should be used mainly for illustrative purposes, since food and fuel energy is not directly interchangeable and since food is grown for many purposes other than mere caloric value, including protein, vitamins, minerals, taste, and aesthetic and cultural appeal.

Energy costs incurred in obtaining essential food nutrients, as well as other substances, also vary with the source. For example, vitamin C occurs abundantly in both oranges (37 mg/100 g) and fresh tomatoes (21 mg/100 g). Although there is almost twice as much vitamin C per unit weight in oranges, about 1 million kcal are required to produce one ton of oranges, whereas only 0.3 million kcal are required to produce a ton of tomatoes. Thus, the same energy investment yields nearly twice as much vitamin C from tomatoes as from oranges. (Tomatoes have the further advantage of being suitable for production in all but the coolest areas.)

With livestock production, significant differences exist among species in the energy-efficiency of producing animal protein. Anywhere from 10 to 90 calories are required to produce one calorie of animal protein, depending on the livestock production system. Milk, eggs, and broiler chickens are products of some of the more energy-efficient livestock systems (Chapters 13 and 14). Livestock products are generally more energy-demanding than plant protein products. A strict vegetarian diet for an adult, for example, requires only about one third as much energy to produce as a high-animal protein diet.

Using grain (or other crops) as animal feed is inherently far more energy-demanding than direct human consumption of grain. As an example, the commercial energy embodied in the production, transportation, and processing of animal feed accounts for one third of such energy input to agriculture in the United Kingdom. As commercial energy becomes more expensive, this type of resource allocation is increasingly difficult to afford, even in industrialized countries; it is out of the question in tropical countries. Since about 10-20 percent of the energy contained in food is available to consumers, the shorter the food chain, the less the amount of energy that is wasted. Animal feedstuffs are less widely used in developing countries, and avoiding them will ease these countries' energy problems. As an alternative, domestic and farm wastes recycled through pigs, chickens, and other small livestock (Chapter 14), or through fishponds (Chapter 15), not only conserve energy by substituting for feedstuffs but also contribute towards a more productive and diversified agriculture. The recognized increased need for draft animals and possibly increased milk production can largely be met through traditional forage, supplemented by suitable protein-rich by-products (such as seed cake left after oil extraction) and by vegetable and root crop wastes, rather than by using grain (Chapter 13).

References

Anderson, R.E. 1979. Biological paths to self-reliance. New York, Reinhold van Nostrand Company, 367 p.

Ashley, H. (ed.) 1976. Energy and the environment: a risk-benefit approach. New York, Pergamon Press, 305 p.

Attiga, A.A. 1979. Global energy transition and the third world. Third World Quarterly I (4):30-56.

Bailey, R. 1979. Development and energy costs: A third world perspective. Third World Quarterly I (4):57-68.

Biswas, A.K. and Biswas, M.R. 1975. Energy and food production. Agro-ecosystems 2:195-210.

Brown, L.R. 1983. Population policies for a new economic era. Washington, D.C., Worldwatch Inst. Paper 53:45 p.

Daly, H.E. and Umana, A.F. (eds.) 1981. Energy, economics, and the environment. Boulder, Colorado, Westview Press, 200 p.

Deudney, D. and Flavin, C. 1983. Renewable energy: the power to choose. New York, Norton, 429 p.

Development Technology Center, Institute of Technology, 1978. A simple solar dryer in Bandung, Indonesia. Appropriate Technology (UK) 5(2):11.

FAO, 1970-1982. State of food and agriculture. Rome, FAO: (various).

FAO, 1973. Energy and protein requirements. Rome, FAO, Food and Nutrition Series 7:118 p.

FAO, 1980. On-farm maize drying and storage in the humid tropics. Rome, FAO, Agric. Serv. Bull.40:40 p.

Flavin, C. 1980. The future of synthetic materials: The petroleum connection. Washington, D.C, Worldwatch Inst. Paper 36:55 p.

Foley, G. and Barnard, G. 1982. Biomass gasification in developing countries. London, Earthscan, 161 p.

Green, M.B. 1978. Eating oil: energy use in food production. Boulder, Colorado, Westview Press, 205 p.

Hill, S.B. and Ramsay, J.A. 1976. Limitations of the energy approach in defining priorities in agriculture. St. Louis, Missouri, Washington Univ., Energy and Agriculture Conf.

Leach, G. 1976. Energy and food production. Guildford, IPC Sci. Tech. Press, 137 p.

Ledec, G. 1983. The political economy of tropical deforestation in Leonard, H.J. (ed.) The Politics of Environment and Development. New York, Holmes and Meier (in press).

Makhijani, A. and Poole, A. 1975. Energy and agriculture in the third world. Cambridge, Mass. Ballinger Publishing Company, 168 p.

McFate, K.L. 1981. Food and energy: challenges and choices. Energy in Agriculture 1:91-98.

Midwest Research Institute, 1978. --- a cash crop for the future? Kansas City, Mo., Midwest Research Institute, 369 p.

National Academy of Sciences. 1976. Energy for rural development: renewable resources and alternative technologies for developing Countries. Washington, D.C., NAS, 306 p.

Pendse, D.R. 1979. The energy crisis and third world options. Third World Quarterly I(4):69-88.

Pimentel, D. and Pimentel, M. 1979. Food, energy and society. London, Arnold, 165 p.

Pimentel, D. (ed.) 1980a. Energy utilization in agriculture. Boca Raton, Fla., CRC Press Handbook Series:496 p.

Pimentel, D. (ed.) 1980b. Food, energy and the future of society. Boulder, Colorado, Westview Press, 390 p.

SERI, 1980. Tree crops for energy coproduction on farms. Solar Energy Research Institute.

Smil, V. and Knowland, W.E. (eds.) 1980. Energy in the developing world: the real energy crisis. Oxford, Oxford Univ. Press, 386 p.

Steinhart, J.S. and Steinhart, C.E. 1974. Energy use in the U.S. food system. Science 184:307-316.

Stout, B.A. and Myers, C.A. 1979. Energy for agriculture. Beltsville Md., USDA., Tech. Info. System: 408 p.

Stout, B. A. et al. 1979. Energy for world agriculture. Rome, FAO Agric. Series 7:286 p.

USDOE 1980 (1979). Soil fertility and soil loss constraints on crop residue removal for energy production. Washington, D.C, Department of Energy's Solar Energy Research Institute, 35 p.

Watanabe, I., Espinas, C.R. Berja, N.S., and Alimagno, E.V. 1977. Utilization of the Azolla-Anabaena complex as a nitrogen fertilizer in rice. Los Banos, Philippines, IRRI Research paper 11:15 p.

Wijewardene, R. 1980. Energy-saving farming systems for the humid tropics. Ibadan, Int. Inst. Trop. Agric. 11(2):47-53.

Wijewardene, R. 1980. Systems and energy in tropical smallholder farming. London, U.K. Commonwealth Secretariat, Appropriate Tillage Workshop: 73-86.

World Bank, 1980. Renewable energy resources in the developing countries. Washington, D.C., World Bank, 33 p.

World Bank, 1981. Mobilizing renewable energy technology in developing countries: Strengthening local capabilities and research. Washington, D.C., World Bank, 52 p.

22
Weed Control

Weeds are, by definition, detrimental to agriculture. They compete with crop plants for light, air, moisture, or nutrients. Weeds also may serve as hosts for plant diseases, insects, and nematodes, and are frequently factors in the epidemic spread of pathogens, virus diseases, and pests. Weed control competes heavily for agricultural labor and capital: agricultural losses due to weeds amounted to over US$1 billion in 1979. On the other hand, some so-called weeds are beneficial under certain circumstances. They provide grazing in areas that cannot be cultivated, or during periods when the land is not producing a crop. They also serve as ground cover, protecting the soil from erosion. Certain weeds provide food during periods of scarcity, while others have medicinal value. Weeds sometimes play a central role in biological pest control by providing food and shelter to the natural enemies of crop pests.

Weed control, therefore, should seek an optimum balance between creating a "floristic desert" and permitting untrammeled weed growth. Clearly, it is unnecessary to invest labor and capital in weed control when weed species are not causing economic damage. Beneficial aspects of plant species commonly labeled weeds need to be recognized; the roles of such plants in providing ground cover, forage, and mulch, as well as in integrated pest management, deserve additional research and exploitation. Weeds are controlled by manual, mechanical, cultural, chemical, or biological methods--all have environmental advantages and disadvantages.

Manual and Mechanical Control: Manual methods of weed control, for both annual and perennial crops, are widely used in the tropics, particularly in areas of abundant labor supply. They are generally effective against annual weed species, but are laborious and inefficient when attempting to control or eradicate vigorous perennial grasses, such as _Imperata cylindrica_ (known in various tropical areas by such names as _lalang, alang-alang_, and _kunai_). Mechanical methods of controlling these grasses are commonly used where large areas are involved. However, vigorous mechanical uprooting of weeds in areas of high rainfall and fragile soil can lead to serious erosion. To put this concern into perspective, the ranking of susceptibility to erosion is led by row crops, the most sensitive. Crops grown on slopes of 3 percent or greater are susceptible, especially if not contour-planted. Monocultures are more susceptible to erosion than strip crops, while rotations are less vulnerable than continuous cropping. Perennial tree crops are least susceptible if the soil is protected by a cover crop; however, after cultivation in young plantings, the broken surface must be covered as rapidly as possible with cover plants to prevent subsequent erosion and re-invasion of weeds.

To minimize crop root damage and soil erosion, tillage should be as infrequent and superficial as possible, and should be adjusted to the minimum depth required to achieve effective weed control. Measures to avert erosion on all sloped land, particularly in high-rainfall areas, are well known; however, in smallholder food cropping areas, such methods are infrequently employed (Chapter 25).

Crop Variety: Hybrids with high seed viability, rapid germination, early emergence, seedling vigor, and rapid growth can favor the crop in competition with weeds simply by surpassing them. However, new hybrids are not necessarily "improved" with respect to weeds; many new varieties, such as the dwarf varieties of wheat and rice, are frequently more vulnerable to weed competition than the tall leafy varieties they replace. While weed control through the use of selected hybrids is environmentally sound to the extent that it does not involve chemicals, the introduction of new food crop varieties has created other, serious, and perhaps long-term agro-environmental problems. These include the loss of germplasm of traditional crop varieties that have evolved over millenia and which serve as the genetic stock for new plant varieties, the increase of genetic uniformity and its concomitant vulnerability to pest and disease attack and unfavorable weather conditions, and the related dependence on costly agricultural inputs to maintain yields (Chapter 24).

Planting Density: One of the traditional methods used for combating weeds is to sow a thick stand of crop. Heavy seeding rates and high crop plant densities give the crop a distinct competitive advantage over the weed population. However, in dry regions, where a relatively low plant population is one of the major means of adjusting field crops to a limited water supply, this method of weed control is self-defeating. Furthermore, while without negative environmental impacts, this method is not practicable when seed is expensive or in short supply.

Mulches: Thick mulches of straw, compost, or such agricultural by-products as coffee bean hulls smother perennial weeds and reduce the germination of annual weeds. Environmentally, mulches are highly preferable because they increase the organic content of soils, recycle nutrients, help retain soil moisture, and prevent erosion. Mulches are not without drawbacks, however. They may provide habitat for some pests (such as slugs and mice) and retard germination by lowering soil temperatures. Nonetheless, mulches merit higher priority than they presently receive. When mulches are managed properly, the farmer does not have to till; this saves energy and labor.

Fire: Controlled burning clears unwanted cover and releases some nutrients into the soils. However, repeated burning without adequate regrowth can exhaust soils by volatilizing sulfur and nitrogen, and by reducing soil organic matter. On balance, the use of fire is often more detrimental than beneficial.

Crop Rotation: A succession of different crops facilitates weed control. Certain weeds are almost obligatory associates of specific crop plants; if these crops are sown continuously, effective weed control may become impossible. In tropical countries, many peasant farmers already practice intensive sequential multi-cropping for this reason. Weeds whose growth and propagation are favored by a specific crop may be weakened by substituting an appropriate alternate crop.

Biological Control: Biological control of weeds by insects has achieved several major successes. One success is the control of over half the Opuntia inermis cactus in Australia by the moth Cactoblastis cactorum. Important potential also exists to employ various plant pathogens for weed control. However, the need to find an insect or other organism that is specific in its choice of host is imperative, because of the serious possibility that a species introduced to control weeds could subsequently endanger crops and other plants of economic value.

Chemical Control: Development of plant growth regulators, selective "hormone" biocides, (e.g., MCPA and 2,4-D) during the 1940's led to significant chemical control methods. Since then the use of contact, translocated, and soil-acting herbicides has proliferated and become a major factor in increasing agricultural productivity. Herbicides account for 60 percent of biocides used in the United States, costing US$3.6 billion annually. The cheaper contact and translocated herbicides, such as paraquat, MSMA, 2,4-D, and dalapon are now commonly used primarily by small rice and tree crop farmers. However, the soil-acting herbicides such as simazine and atrazine have largely remained within the domain of major maize and sugar cane growers only. Pimentel (1981) argues that elimination of herbicides might require an increase of as much as 46 percent in agricultural land area to yield similar quantities of food and fiber. Elimination of herbicides could also reduce farm revenues by as much as 31 percent, resulting in worldwide economic losses of US$13 billion (Pimentel, 1981).

With weeds posing possibly the greatest single limitation on farmers' productivity in the humid tropics, chemical controls are likely to play an increasing role, particularly in those areas where labor is in short supply. Investigations have shown that in many small farm areas, a combination of manual weed control and applications of herbicides at low rates is most effective. However, it is important to recognize the dynamic nature of weed populations that affect biocidal effectiveness, and the social or environmental problems that biocides often present.

Weeds have not yet developed resistance to herbicides to the degree that insects have to insecticides. However, when one biocide is used continuously on a crop it is inevitable that, sooner or later, the weed population will change, with tolerant, mostly perennial, species dominating as susceptible species are locally eliminated. Also, the differential response of biotypes within one species may result in the development of a tolerant population of that species. Weed species shifts or secondary pest outbreaks are a frequent result of repeated biocide usage. For example, Phalaris spp. weeds have increased rapidly in 2,4-D treated wheat in North India. Rottboellia exaltata has become a serious problem in maize, upland rice, and sugar cane cultivation, mainly in response to atrazine application. Extensive use of paraquat in banana, citrus fruit, and coffee cultivation has led to the dominance of Parthenium hysterophorus in Cuba (Plucknett, 1977).

Herbicides often present other environmental problems. Erosion on hilly slopes may result from complete vegegation removal following initial herbicide application. Soil-acting biocides, such as trifluralin, may harm fish life if they wash into waterways. Crops may be damaged by direct contact with the immediate spray (or spray droplets that may travel considerable distances under certain conditions). Through a variety of mechanisms, herbicides may also encourage insect and pathogen attacks on crops.

There is also a risk of direct harm to spray operators, particularly during the handling of concentrates and preparation of spray solutions. The hot, humid conditions under which spraying is generally carried out also increases the risk of skin uptake of the chemical, followed by dermatitis or other illness. The most notorious example of toxicity has concerned the herbicide 2,4,5-T, used as a tree killer and for the control of shrubs and creepers. This biocide itself is not unusually harmful, but it often contains harmful levels of the impurity dioxin, which is extremely toxic and is known to cause birth defects, genetic mutation, and chronic ailments (Du and Young, 1980). In modern formulations, this impurity can be reduced or even eliminated; only such safer products are recommended for use.

Where biocide use is deemed essential because of labor constraints, following health and safety precautions for workers avoids most of the possible damage from misuse. For example, concentrates should be stored separately from foodstuffs; spray solution should be prepared only by properly-instructed workers, preferably wearing goggles and rubber gloves; spraying equipment should be free from leaks; and workers should wear some form of protective clothing to minimize contact with the spray. Washing of clothes and the individual after spraying further reduces potential health problems. To increase safety (and effectiveness), biocides should be applied during the weed's most vulnerable growth stage and under conditions where the risk of spray drifting onto the crop or adjacent areas is minimal. Avoiding contamination of irrigation water also reduces health risks. Those biocides causing serious health effects are banned in some countries; suitable alternatives are available.

Eventually, biocides in soils may be destroyed by micro-organisms or by chemical or photo-decomposition, inactiviated through absorption in soil colloids, or dissipated by leaching or volatization. Usually, the principal factor in the degradation of biocides is biological. Decomposition of biocides to sub-toxic levels by soil organisms may require a few days or up to a year or more. In the humid tropics, this decomposition is much more rapid than in drier or cooler regions. In general, well-managed irrigated soils are warm, moist, fertile, and well-aerated; all these conditions are conducive to relatively rapid biocide decompositon. By contrast, in unirrigated agriculture during extended periods of drought, microbiological decomposition of biocides may cease. Microbial decomposition of biocides may also be impeded by certain of the biocides themselves, including DDT, benzene hexachloride, and chlordane, all of which can accumulate in soils and are toxic to some micro-organisms. The adverse consequences of biocides for weed control can be minimized if appropriate precautions are taken and if primary emphasis is placed on manual, mechanical, cultural, and biological control methods.

References

Altieri, M.A., Schoonhoen, A. van, and Doll, J. 1977. The ecological role of weeds in insect pest management systems: a review illustrated by bean (Phaseolus vulgaris) cropping systems. PANS 23(2):195-205.

Baloch, G.M. 1977. Insects as biological control agents of field bindweed, Convolvulus arvensis. PANS 23(1):58-64.

Buckley, N.G. 1980. No-tillage weed control in the tropics. Haslemere, U.K., ICI Plant Prot. Div:12-21.

Cherrett, J.M. and Sagar, G.R. (eds.) 1976. Origins of pest, parasite, disease and weed problems. Oxford, Blackwell Scientific Publications, 413 p.

Crafts, A.S. 1975. Modern weed control. Berkeley, California. University of California Press, 440 p.

Day, B.E. (ed). 1968. Weed control. Washington, D.C., National Academy of Sciences, 471 p.

Du, J. and Young, P. J. 1980. Agent orange: the bitter harvest. Sydney, Hodder and Stoughton, 285 p.

Fryer, J.D. and Evans, S.A. (eds.) 1968. Weed control handbook. Oxford, Blackwell Scientific Publications, 2 vols.

Litsinger, J.A. and Moody, K. 1976. Integrated pest management in multiple cropping systems. (293-316) in Papendick, R.I., Sanchez P.A. and Triplett, G.B. (eds.) Multiple cropping. Madison, Amer. Soc. Agron., 378 p.

Meister Publishing Company, 1976. Farm chemicals handbook. Willoughby, Ohio.(v.p.)

Midwest Research Institute, 1976. Assessment of ecological effects of extensive or repeated use of herbicides. Kansas City, Mo., Midwest Research Institute, 369 p.

Moody, K. 1975. Weeds and shifting cultivation. PANS 21(2):188-194.

Oka, I.N. and Pimentel, D. H. 1976. Herbicide (2,4-D) increases insect and pathogen pests in corn. Science 193:239-240.

Pimentel, D.H., et al. 1974. Herbicide report. Chemistry and analysis. Environmental effects. Agriculture and applied uses. Washington, D.C., Hazardous Materials Advisory Committee, Environmental Protection Agency.

Pimentel, D.H. 1981. Handbook of pest management. Boca Raton, CRC Press, 3 vols.

Plucknett, D.L.., et al. 1977. Approaches to weed control in
     cropping systems (294-308) in Cropping Systems. Los Banos,
     Philippines, IRRI symposium, 454 p.

Shaw, W.C. 1976. Weed science - revolution in progress.
     Northeastern Weed Science Society 30:402 p.

Tabora, P.C. 1979. Weed control in Abaca in the Philippines,
     (164-168) in Weed Control in Tropical Crops. Manila,
     Philippines, Weed Sci. Soc. Philippines Inc.

Thomson, W.T. 1977. Herbicides. Fresno, Ca., Thomson Publications,
     281 p.

Versteeg, M.N. and Maldonado, D. 1978. Increased profitability
     using low doses of herbicide with supplementary weeding in
     smallholdings, Coperhalta project, Peru. PANS (Pest Articles
     and News Summaries) 24:327-332.

Wapshere, A.J. 1975. A protocol for programmes for biological
     control of weeds. PANS (Pest Articles and News Summaries)
     21(3):295-303.

Williams, R.D. 1979. Weed management in vegetable crops. (149-163)
     in Proc. Symposium on Weed Control in Tropical Crops. Manila,
     Weed Sci. Soc. Philippines Inc.

23
Water Weed Control

Proliferation of weeds in waterways interferes with irrigation, navigation, energy generation, and insect and disease control. Such proliferation also impairs water quality and fisheries. Some control methods, such as widespread spraying of biocides, can create additional environmental disruption. Other methods link control with productive use by exploiting water weeds for economic benefit, while controlling growth and preventing economic loss. In environmental terms, natural, manual, and mechanical control methods are preferred over the use of chemicals. The environmental costs and benefits of the various water weed control measures are adequately described in the listed references, and are therefore largely not repeated here.

The use of aquatic weeds contributes markedly to their control while generating valuable products: meat, eggs, fish, edible vegetation, fertilizer, energy, paper pulp, and waste processing. Use of aquatic weeds returns nutrients to human and agricultural systems. Where utilized for energy and pulp and paper, water weeds can slow the drain on less rapidly renewable resources such as timber. Combining control methods is frequently the best way to benefit from water weeds, while still maintaining the water body's value for fishing, navigation, and other uses.

Chemical Control: Chemical control methods, though costly, have been relatively effective in killing water weeds in tropical areas. Water hyacinth infestation of Brokopondo Reservoir in Suriname was combated by a US$2.5 million 2-4,D aerial spraying program during the period from 1964 to 1970. However, severe environmental costs resulted from the spraying. The anaerobic rotting of target plants severely depleted aquatic oxygen and produced gases that destroyed fish and precluded many other uses of the reservoir (Gijzel, 1977). Reduced fish stocks caused malnutrition among people dependent on the river for fish (Panday, 1977). In retrospect, the weed problem was not adequately assessed before spraying, and the spraying itself may have compounded the negative results.

Deoxygenation is a serious problem in itself. Aquatic plants characteristically produce masses of vegetation: some water weeds are, in fact, among the most productive plants on earth. For example, water hyacinth stands weighing 470 tons/ha often grow at rates of roughly 4.8 percent per day; growth rates of 800 kg of dry matter/ha/day have been measured. When killed by biocides, this voluminous biomass sinks and decays, causing varying degrees of deoxygenation. This problem is particularly serious for people who depend directly on the affected waters. Control methods that involve removal of aquatic weeds, either mechanically or by conversion to protein by manatees, fish, or other herbivores are environmentally preferable to methods that leave the weeds to decay on site.

Manual and Mechanical Control: Since aquatic weeds consist of 85-95 percent water, harvesting the heavy, intertwined, and sometimes bottom-rooted plants is energy-expensive. Mechanical clearing is simply not practicable for large expanses, such as the 1,000 km^2 covered with **Salivinia molesta** on the 4,300 km^2 Kariba reservoir in Zambia. Manual harvesting is useful for small-scale collecting of mulch or feed for animals, but is even less feasible for control over large areas.

After aquatic weeds are harvested, mechanical de-watering with presses is quite energy efficient. In a few hours, this process can yield tons of vegetable matter whose reduced water content is comparable to that of terrestrial forage grasses. One small (250 kg) screw press designed at the University of Florida can remove as much as 50 percent of the water content of up to 4 tons of chopped water hyacinth per hour. The press can be carried by truck, trailer, or barge to remote locations (NAS, 1976). Water squeezed out of aquatic plants can be returned to waterways without polluting. Thus, while large-scale manual or mechanical clearing of water weeds is usually not practicable, localized harvesting and de-watering for mulch, feed, and other uses are both feasible and environmentally desirable.

Natural Control Options: Biological control (Chapter 18) has occasionally been spectacularly successful for water weeds. For example, **Salvinia** in tropical Australia is successfully controlled by a small weevil (**Cyrtobagous singularis**) from Brazil (Room, **et al.**, 1981). Grass carp and other herbivorous fish, manatees, crayfish, ducks, geese, swans, and other herbivores have all been shown effective in checking the growth of certain kinds of water weeds (but not the most troublesome species), while often providing a source of food. Chinese Grass Carp (**Ctenopharyngodon idella**) are being used in India's Chambal (Rajasthan) project. In Arkansas, from 1970 to 1975, this carp cleared 20,000 hectares of weed-infested lakes. A fast-growing species that can reach 32 kg, this carp is described by the United States National Academy of Sciences as an "exceptionally effective control agent for submerged weeds." Chinese Grass Carp yields of 1,500 kg/ha have been obtained in tropical weed-infested waters, usually in polycultures involving several different fish species. Various species of **Tilapia**, two species of **Saratherodon**, as well as **Metynnis roosevelti**, **Mylossoma argenteum**, Silver Carp (**Hyphothalmichthys molitrix**), and Common Carp

(Cyprinus carpio) are among many other herbivorous fish species that have been successfully used for weed control in irrigation channels and large water bodies. Crayfish of the genera Orconectes, Procambarus, and Cambarus also offer significant potential for weed control.

Introducing non-native fish or other aquatic species involves the risk of seriously disrupting native aquatic life. For example, introducing carp may damage native fish because the carp might increase the water's turbidity and eat the eggs or fry of native species. Aggressive non-native species sometimes even outcompete native species to the point of causing their extinction. Because they are completely irreversible and harm the interests of present and future human generations, human-induced species extinctions should be explicitly avoided in project design and implementation (Chapter 24). For water weed control and freshwater fisheries (Chapter 15) projects, non-native species should not be introduced until a scientific assessment is made of the biological risks of the introduction. Also, using native species (whenever practicable) is much less risky than the introduction of exotic species.

Manatees are large, herbivorous aquatic mammals. They can consume as much as a twentieth of their body weight each day--approximately 20 kg of wet vegetation. These creatures are adaptable to confinement, as well as to waters that are fresh, brackish or saline; acidic or alkaline; muddy or clear. Manatees have kept some of the water systems of Georgetown, Guyana relatively clear of weeds for over 25 years. However, the manatee is an endangered species--easy to catch, kill, and eat. Manatees breed very slowly, and nowhere are populations large enough to support an export trade to nations that would like to introduce them into weed-clogged waterways.

Ducks, geese, and swans eat considerable quantities of vegetation and have long been effective in controlling weeds, although their role as such has not been emphasized. Five to eight Muscovy Ducks per hectare will keep duckweed and some other small aquatic plants under control. Use of ducks and other water fowl in waste water/fertilizer systems is well-established and highly successful in Asia, particularly China. In these systems, the birds feed on water weeds fertilized by organic wastes; they also control insects and the snail vectors of schistosomiasis. Other herbivorous or omnivorous animals, such as water buffalo, capybara, and pigs will also consume water weeds that either are growing in the water or have been harvested as fodder.

Use as Fertilizer: Water weeds can serve as a green mulch, whether on the soil surface as a simple mulch, plowed under, or buried in troughs. Water weeds can also be composted with soil, ash, and a small amount of animal manure. For example, the Home Gardens Division of the Sri Lanka Department of Agriculture combines chopped water hyacinth and city refuse, together with small quantities of ash, earth, and cow manure, to generate 80 tons of compost per month (NAS, 1976).

Use as Animal Feed: Most aquatic weeds contain 10-26 percent crude protein (dry matter basis), which equals or is higher than the content of most forages. Fibrous waterweeds, such as reeds and cattails, are potential substitutes for other roughages eaten by ruminant animals. NAS (1976) reports that:

"In Southeast Asia, some nonruminant animals are fed rations containing water hyacinth. In China, pig farmers boil chopped water hyacinth with vegetable wastes, rice bran, copra cake, and salt to make a suitable feed. In Malaysia, fresh water hyacinth is cooked with rice bran and fish meal and mixed with copra meal as feed for pigs, ducks, and pond fish. Similar practices are much used in Indonesia, the Philippines, and Thailand."

High moisture and mineral content are limiting factors in the use of aquatic plants for animal feed. The first limitation is corrected by dewatering or ensiling, which produces a palatable, digestable, and nutritious silage that is readily accepted by cattle and sheep. Mineral content of some aquatic plants can be so high, however, that they cannot be used as fodder. Similarly, water weeds should never comprise the entire diet of livestock, because of their excessive mineral content. When harvested at the right stage of growth, however, succulent young water weeds are relatively low in minerals and highly suitable for use as supplement fodder.

Use as Fiber and Food: Reeds, bulrushes, cattails, and papyrus are prominent features of many marshes and swamps. While they provide essential habitat for many wildlife species (particularly birds), their proliferation can hinder agriculture, fishing, recreation, and navigation. Harvested for centuries, these fibrous plants have diverse uses, including pulp and paper production. These plants can readily be encouraged to grow in appropriate water bodies. For example, the Romanian government harvests 125,000 tons per year of reeds from the Danube River delta. The potential of water weeds for fiber production is likely to become even more attractive as petroleum-based synthetics increase in cost.

Water spinach, watercress, floating and wild rice, chinese water chestnut, taro, and arrowhead are all useful, food-producing aquatic plants. Floating vegetable gardens, made from bottom muck that is laid on mats of aquatic weeds, are common in Burma, Bangladesh, Kashmir, and parts of Central America.

Cattails, while serious nuisances when they invade rice fields, farm ponds, lakes, and canals, are generally useful and productive plants. Almost all parts of the plant are edible. The root-like rhizomes contain as much protein as corn or rice and more carbohydrates than potato. A single hectare of these plants may yield over 7,000 kg of rhizomes. NAS (1976) states that "perhaps the cattail's greatest economic potential is as a source of pulp, paper, and fiber." Recent studies in Mexico show that woven cattail leaves, coated with plastic resins, have potential as building siding and roof tiles. The resulting product is said to be at least as strong as fiberglass. Other aquatic weeds offer similar potential.

Use as Energy: Water hyacinths can be converted into a methane-rich gas at the rate of approximately 70,000 cubic meters of biogas per hectare. NAS (1976) calculated that each kg (dry weight) of water hyacinth yields up to 370 liters of biogas, with an average methane content of 69 percent and a calorific heating value, when used as a fuel, of about 22,000 kJ per cubic meter (895 BTU per cubic foot). Of the many thousands of biogas-generating digesters operating throughout the world, most are in China, India, Taiwan, and Korea. Considerable potential exists for the use of water weeds in these (and other) digesters, depending upon the costs of cutting and transporting the weeds.

Water Treatment Using Aquatic Weeds: Many aquatic weeds absorb inorganic and some organic compounds from water and incorporate the materials within their own structure. In effect, they thereby serve as scavengers. Living plants strip effluent of pollutants (i.e., nutrients and metallic minerals, including heavy metals), thus reducing the environmental damage that results when effluent is released into waterways.

Plants grown on waste water can also be harvested and utilized. This process is environmentally useful, economically advantageous, and simple to manage. For example, in rural Vietnam, barns are built so that manure from sheep and goats inside the structure falls into a pond to fertilize a crop of aquatic plants. The plants, in turn, are harvested regularly and provide feed for the animals. On a larger scale, sewage from the city of Lucedale, Mississippi is lagooned and then pumped into a series of channels. The area's subtropical climate and the waste water's rich nutrients result in huge crops of water hyacinth. Designed to help the water hyacinth extract maximal quantities of nutrients, the one meter-deep channels also facilitate harvest of the plants. After harvesting, the vegetation is processed into animal feed, fuel, and fertilizer. Because water weeds accumulate heavy metals and other toxins in their tissues, it is important to test sewage water for concentrations of such substances, particularly if the plants are to be used as animal feed.

References

Ahmad, N. 1977. Water-hyacinth and fish farming. Pakistan J. Science. 29:55-59.

Andres, L.A. and Bennett, F.D. 1975. Biological control of aquatic weeds. Ann.Rev. Entomol. 20:31-46.

Biswas, D.K. 1978. Integrated measures for control and utilization of aquatic weeds. Bhagirath (India) 25(3):138-142.

Dawson, J. 1978. Aquatic plant management in semi-natural streams: the role of marginal vegetation. J. Env. Management 6(3):213-221.

Druijff, A.H. 1973. Possible approaches to water hyacinth control and utilization of aquatic weeds. Bhagirath (India) 25(3):138-142.

Freeman, T.E., Zettler, F.W. and Chandattan, R. 1974. Phytopathogens as bio-controls for aquatic weeds. PANS 20(2):181-184.

Fryer, J.D. and Evans, S.A. (eds.) 1968. Weed control handbook. Oxford, Blackwell Scientific Publications, 2 vols.

Gijzel, W.P. 1977. The effects of some biocides on the ecosystem of a man-made lake (39-40) in Panday, R.S. (ed.) Man-Made Lakes and Human Health. Paramaribo, University of Suriname, 73 p.

Mitchell, D.S. and Thomas, P.A. 1972. Ecology of waterweeds in the neo-tropics: an ecological survey of the aquatic weeds Eichhornia crassipes and Salvinia species, and their natural enemies in the neotropics. Paris, Unesco, Technical papers in Hydrology 12:50 p.

Monsod, G.G. 1979. Man and the water hyacinth. New York, Vantage Press, 48 p.

NAS, 1976. Making aquatic weeds useful: Some perspectives for developing countries. Washington, D.C., National Academy of Sciences, 174 p.

Panday, R.S. (ed.) 1977. Man-made lakes and human health. Parimaribo, University of Suriname, 73 p.

Perkins, B.D. 1974. Arthropods that stress water-hyacinth. PANS 20(23):304-314.

Pieterse, A.H. 1978. The water-hyacinth (Eichhornia crassipes): a review. Abstracts on Tropical Agriculture 4(2):1-42.

Pieterse, A.H. 1979. Aquatic weed control in tropical and sub-tropical regions (130-136) in Beshir, M. E. and Koch, W. (eds.) Weed Research in the Sudan. Hohenheim Univ. vol. I.

Room, P.M., Harley, K.L.S., Forno, I.W. and Sands, D.P.A. 1981. Successful biological control of the floating weed Salvinia. Nature 294:78-80.

Spencer, N.R. 1974. Insect enemies of aquatic weeds. PANS 20(4):444-450.

24
Conserving Biological and Genetic Diversity

"The worst thing that can happen--will happen--in the 1980's is not energy depletion, economic collapse, limited nuclear war, or conquest by a totalitarian government. As terrible as these catastrophes would be for us, they can be repaired within a few generations. The one process ongoing in the 1980's that will take millions of years to correct is the loss of genetic and species diversity by the destruction of natural habitats. This is the folly our descendants are least likely to forgive us".

--E. O. Wilson

The world's biological diversity is dwindling as destruction of habitats, particularly tropical forests, continues at rapid rates. Moreover, the genetic diversity of vital food crops is also being eroded as a relatively few genetically uniform seed and crop types (e.g., "High Yield Varieties," HYVs) replace the one-time multiplicity of indigenous and traditional varieties. Germplasm conservation measures mitigate these ominous trends.

Genetic impoverishment through extinction of plants and animals is commonly thought of in association with crop species and endangered exotic bird or mammal species. However, habitat destruction imposes a loss of species in all taxonomic groups and is, as yet, largely unexplored and unappreciated by humans. Estimates place the world's stock of biological species at 10 million, of which an estimated 8.5 million still remain to be discovered and named! The potential contributions of these unknown species to human welfare through agriculture, medicine, pharmaceuticals, and industry--as well as the actual contributions of these species to maintenance of global ecological stability--defy calculation. The immensity of the problem, especially in the tropics, deserves recognition and accommodation in all development schemes involving habitat alteration, if long-term enviromental risks are to be reduced.

Development strategies involving expansion into new and often agriculturally marginal lands contribute directly to the loss of genetic diversity. Rising demands of industrialized nations, and affluent sectors of developing nations, for products such as tropical hardwoods and beef contribute to loss of natural habitats, particularly species-rich tropical forests. Strategies that substitute monocultures in place of diversified traditional

- 208 -

Figure 24.1: Ranking of Genetic Preservation Options

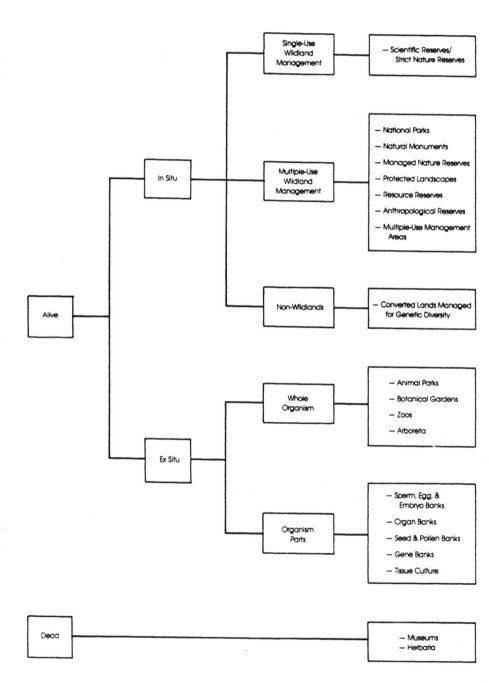

World Bank—25102

agricultural patterns can threaten the genetic diversity of basic food crops. Thus, while species inventories, national parks and other protected areas, and germplasm collections are valuable and necessary measures, genetic loss relates even more closely to the economic and political pressures inherent in certain development strategies.

Development projects should be designed and implemented in ways that prevent or minimize genetic losses. Individual development projects can address the genetic threat of loss through such measures as species inventories, germplasm collection, and setting aside biotic reserves (as noted in Chapter 17 and Figure 24.1).

The United States National Academy of Sciences (1978) has identified several categories of particular concern regarding genetic diversity: export crops of developing countries, forest germplasm, drug plants, livestock germplasm, pests and pathogens, marine invertebrates, and economically important micro-organisms. Several of these groups relate to agricultural development projects.

Basic Food Crops: The Consultative Group on International Agricultural Research (CGIAR), an organization which supports international agricultural research centers, addressed the genetic diversity problem several years ago by forming the International Board for Plant Genetic Resources (IBPGR). The IBPGR has stimulated national programs and provided funds to establish world collections of endangered germplasm. These collections are an encouraging beginning for the preservation of threatened gerplasm. Much remains to be done, however, especially with crops that are not today of major export or commercial interest (Toll, et al., 1980).

No techniques yet exist for the off-site (ex situ) conservation of vegetatively propagated plants, other than the costly maintenance of nursery-grown stocks. Stocks of other important crop plants remain uncollected. For example, while United States and Canadian collections maintain strains of such basic staple crops as wheat, maize, and barley, vegetables continue to be virtually unrepresented in germplasm collections. Even in such collections, genetic diversity can be lost as techniques for mass rearing are adopted and culture conditions become increasingly uniform through artificial control. Discarding all strains except those with characteristics suited to current methods of domestic production and present-day conditions is undesirable, in that it forecloses future options. The loss of genetic diversity poses a serious threat as environmental conditions change, as pests or parasites develop or shift their susceptibility to predators, or as needs for different sets of characteristics arise.

When seeds of new crop strains are introduced, the indigenous varieties should be conserved. If no formal facilities exist, samples of the original varieties can be maintained in "museum strips" using methods of cultivation as similar as possible to the original. In the short run, this will provide insurance in case the new varieties fail. In the long run, it will maintain genetic diversity to preserve future options and reduce risks.

Tropical Export Crops: The United States National Academy of Sciences (1978) has stated that tea, coffee, rubber, oil palm, cacao, and other tropical crops are suffering reduction of their genetic base "through the planting of vast areas with single types, such that the enormous native diversity initially characteristic of these species is being rapidly destroyed." Rubber, for example, cannot be planted profitably and extensively in the continuously wet areas of New World because of the lethal leaf blight fungus. Inadvertant introduction of this fungus into continuously wet Asian rubber-producing areas could virtually destroy a crop which is of increasing value as petroleum-based synthetic rubber becomes more expensive. Gene pools of diverse rubber varieties are, therefore, essential to respond to likely future breeding needs. The rubber industry has, in fact, taken useful steps in this direction. All producing countries maintain diverse planting material of the various **Hevea** species and a quarantine station has been established in Trinidad to facilitate the movement of proven disease-free material between countries. This situation is mirrored to a lesser extent with other crops, but there can be no complacency. Crops propagated vegetatively (e.g., bananas) are the most liable to become genetically diminished. Some private institutions are maintaining seed libraries of traditional varieties and encouraging local growers to undertake genetic diversification (HDRA, 1979).

Forest Germplasm: The increasing loss of tropical forests--believed to contain more species than any other ecosystem--is now recognized as a major world problem (Chapter 12). National governments and development agencies are becoming more aware of the importance of tropical forests. The importance of forest cover to watershed management, long understood by environmentalists, is only just being accommodated by development planners. Uncontrolled forest clearing from poorly designed development projects has had disastrous effects in the past, including erosion, sedimentation, and flooding.

The preservation of biological diversity in other protected areas should be a routine component of many types of agricultural development projects. For example, the Dumoga National Park in Sulawesi, Indonesia protects the watershed of a major irrigated rice project, thereby improving water control and minimizing siltation of irrigation canals. It also preserves much of the rich fauna and flora that are unique to Sulawesi.

Drug Plants: Between 1959 and 1973, plant-derived drugs accounted for about 25 percent of all prescriptions dispensed from pharmacies in industrial countries. In 1974, the retail value of such drugs had reached approximately US$3 billion in the United States alone (Prescott-Allen, 1982). Indigenous or tribal societies have long utilized native plants therapeutically; a "self-interest" argument for the preservation of these peoples and their traditional ways is that industrialized societies will gain new medicines (World Bank, 1982). The United States National Academy of Sciences summed up the situation by stating:

"It appears that an intensified last look, so to speak, at folk medicines should be undertaken before these primitive societies wholly disappear, lest some effective remedies already identified in their cultures be needlessly overlooked" (NAS, 1978).

Species of pharmaceutical value remain to be inventoried, including those containing antibacterial, antiviral, and tumor-inhibiting substances, as well as anticoagulants and neurobiologically active materials. Many of these substances are found in species of the highly diverse tropical ecosystems which therefore provide the greatest potential source of as yet undiscovered compounds.

Livestock Germplasm: Changes in environmental and economic circumstances, as well as changing husbandry practices, intensify the need to conserve livestock germplasm. One such need is the shift from livestock species fed largely on grain to those fed on non-grain feedstuffs, as grain becomes too costly, partly in response to human food requirements. Wild relatives of livestock have unique genetic characteristics not present in domestic strains, such as high fertility and milk production in sheep, high growth rates in goats, and high productivity in rabbits and other species. The United Nations Man and the Biosphere (MAB) Program has stated that "generally speaking, the wild relatives of domestic animals are in a critical condition and the survival of many is seriously endangered. For example, the wild relatives of the horse, ass, cow, sheep, buffalo, goat, yak, are all either endangered as species or have endangered races. In many instances, populations have been reduced by habitat destruction combined with excessive exploitation. In addition, interbreeding with domestic livestock is a threat to the genetic integrity of many wild populations."

The worldwide protection of game animals, particularly in Africa, is a related concern. The underutilized potential of game for protein production, especially in tsetse-infested areas, is mentioned in Chapters 14 and 19. Habitat protection to conserve genetically diverse and agriculturally important insect species also deserves high priority. Particularly important in economic terms are bees and other pollinators, and species which prey upon agricultural pests or disease vectors.

Fishes and Marine Invertebrates: Marine fishes and, to a lesser degree, freshwater fishes, are among the least-known groups of vertebrates, despite their importance as a food source. Most undescribed fishes are marine, although areas such as the Amazon Basin also contain many undescribed species. Entire species of commercially important fish have been lost locally, usually by overfishing, aggravated by man-made changes in the physical or biological environment. Marine fishery projects should promote controlled exploitation of several fish species to avoid overfishing of any one species. Mangrove stands and other breeding areas important for fish production (Chapter 16) are vulnerable to destruction by poorly-planned development activities. Projects should also be designed to minimize pollution potentially harmful to freshwater and marine fish species. Genetic research on lobsters, shrimp, crabs, clams, scallops, oysters, and other marine invertebrates has been minimal, notwithstanding the large and increasing economic importance of these organisms. Protection of their breeding grounds and investigation and maintenance of their genetic diversity should receive high priority.

Natural Habitat Conservation: Development projects should be designed and implemented with regard to preventing or minimizing species extinctions. Projects that convert tropical forests, for example, are environmentally and (in the long term) economically more destructive to the extent that they cause the loss of genetic resources. Rational management of utilizable native plant and animal species in some forest and savanna areas can almost certainly lead to higher human carrying capacities than clearing natural vegetation on characteristically poor soils, without causing genetic depletion.

Biologists now realize that the most diverse and therefore species-rich ecosystems are precisely those which can be most vulnerable when disturbed. The tropical rain forest, for example, is such a complex matrix of interrelating animals, plants, and micro-organisms that if removed on a large scale, regeneration of the original forest is virtually impossible. With this insight, designation of conservation areas--particularly in tropical forests, wetlands, coral reefs, coastal zones, and mangrove swamps--can minimize current species losses. Such designations recognize that the only practical means to preserve the germplasm of unidentified organisms, as well as identified species of unassessed value, is through the preservation of habitats extensive enough reasonably to assure their long-term viability.

Conservation of wildlands in the design of development projects has yet to be made routine. Some tourism projects have achieved significant conservation gains, thereby preserving genetic resources. For example, several coastal tourism projects located in the Caribbean and the Indian Ocean regions feature protection of coral reefs and other tropical shallow-water ecosystems from human damage, such as the removal of coral and shells for souvenirs, spearfishing, cutting of the coral reefs for limestone, and freshwater storm and sewage drainage (which causes osmotic stress to coral and other reef organisms.) Other tourism projects support protection of wildlife and national parks. For example, a project in Kenya provides for improvements to one national park and three game reserves, as well as training for wildlife and fisheries management and the establishment of anti-poaching units. A project in Tanzania includes anti-poaching equipment, while a project component in the Ivory Coast consists of a resettlement program for squatter families who periodically invade a nature reserve.

Projects in ecologically unique or significant areas require extreme caution if irreversible environmental damage and major biotic losses are to be avoided. UNESCO's Man and the Biosphere (MAB) Program started in 1973 states that the need for a carefully selected network of protected areas is most acute "where forms of land use destructive to natural communities are likely to take place in all areas other than those set aside and specially protected, or where uncontrolled exploitation or destruction of indigenous species is likely to take place wherever those species are not specially protected."

In protected area design, it is desirable to include examples of different ecosystems within one large protected area; to choose tracts rich in species, including endemics or relics; and to include breeding, feeding or staging sites of migratory animals. Also, where consistent with pest management objectives, agricultural patterns can be planned with margins of undisturbed vegetation to provide significant reserves of natural habitat. Hedgerows, terraces, stone fences and windbreaks for example are important wildlife reserves in otherwise fully developed landscapes throughout the world.

The MAB Program has stressed, however, that "reserves are unlikely to continue to serve their designated purpose unless some degree of planning and control of land and resource use outside their boundaries can be accomplished." At least one reserve set up in association with a development project failed when it was invaded by settlers, in part because land holdings were extremely inequitable in the area surrounding the reserve. Projects and policies which successfully address such basic distributional issues will reduce the need for colonization of species-rich wildlands, thereby protecting our global genetic heritage for posterity.

References

Allen, R. (ed.) 1980. How to save the world: strategy for world conservation. Gland, IUCN, UNEP, WWF:150 p.

Anon, 1978. Proceedings of the workshop on the genetic conservation of rice. IRRI, Los Banos, Philippines:60 p.

Carp, E. (ed.) 1972. Conservation of wetlands and water fowl. Slimbridge, International Wildlife Research Bureau.

Consultative Group on International Agricultural Research Secretariat. 1979. Report on the Consultative Group and the International Agricultural Research System. An Integrative Report. Washington, D.C., World Bank, 60 p.

Dasmann, R.F. 1972a. Planet in peril? Man and the biosphere today. New York, World Publishers, 242 p.

Dasmann, R.F. 1972b. Towards a system for classifying natural regions of the world and their representation by natural parks and reserves. Biological Conservation 4:247-255.

Dasmann, R.F. 1973a. Classification and use of protected natural and cultural areas. Morges, IUCN Occasional Paper 4:24 p.

Dasmann, R.F. 1973 b. A system for defining and classifying natural regions for purposes of conservation: a progress report. Morges, IUCN Occasional Paper 7:47 p.

Dasmann, R. F. 1984. Environmental conservation. New York, Wiley, 450 p.

Duffey, E. and Watt, A.S. (eds.) 1971. The scientific management of animal and plant communities for conservation. Oxford, Blackwell, 652 p.

Eckholm, E. 1978. Disappearing species: the social challenge. Washington, D.C, Worldwatch Paper 22:38 p.

Ehrlich, P. R. and Ehrlich, A. H. 1981. Extinction. New York, Random House, 305 p.

Elliott, H.F. (ed.) 1974. Second World Conference on National Parks. Morges, IUCN, 504 p.

FAO, 1975. The methodology of conservation of forest genetic resources: report on a pilot study. Rome, FAO:127 p.

Frankel, O.H. (ed.). 1973. Survey of crop genetic resources in their centres of diversity. Rome, FAO-IBP:164 p.

Frankel, O.H. 1981. Conservation and evolution. New York, Cambridge Univ. Press, 317 p.

Frankel, O.H. and Bennett, E. 1970. Genetic resources in plants; their exploration and conservation. Oxford, Blackwell, IBP Handbook No. 11:554 p.

Hawks, J.G. 1983. The diversity of crop plants. London, Harvard Univ. Press, 208 p.

HDRA, 1979. The world's vanishing vegetables. Henry Doubleday Research Association Newsletter 79:29-31.

Hooper, M.D. 1971. The size and surroundings of nature reserves (555-561) in Duffey, E. and Watt, A.S. (eds.). The scientific management of animal and plant communities for conservation. Oxford, Blackwell, 652 p.

IUCN, 1970. Creative conservation in an agrarian society. Morges, IUCN 21:98 p.

IUCN, 1973. 1973 United Nations List of National Parks and Equivalent Reserves. Gland, Switzerland, IUCN, 601 p.

IUCN, 1982. The world's greatest natural areas. Gland, Switzerland, IUCN.

Luther, H. and Rzoska, J. 1971. Project Aqua: a source book of inland waters proposed for conservation. Oxford, Blackwell, IBP Handbook 21:239 p.

Man and the Biosphere Program Report Series: Green Books (Sept. 1973). Conservation of natural areas and of the genetic material they contain. Morges, Switzerland.

NAS, 1978. Conservation of germplasm resources: an imperative. Washington, D.C., National Academy of Sciences, Institute of Medicine, National Academy of Engineering, 118 p.

Nicholson, E.M. 1968. Handbook of the conservation section of the International Biological Programme. Oxford, Blackwell, IBP Handbook 5:84 p.

Prescott-Allen, R. & C. 1982. What's wildlife worth? Economic contributions of wild plants and animals to developing countries. London, Earthscan, 92 p.

Radford, G.L. and Pankhurst, R.J. 1973. A conservation data base. New Phytologist 72:1191-1206.

Roche, L. 1971. The conservation of forest gene resources in Canada. Forestry Chronicle 47:215-217.

Soule, M. E. and Wilcox, B. A. (eds.) 1980. Conservation biology: an evolutionary-ecological perspective. Sunderland, Massachusetts, Sinauer, 395 p.

Sykes, J.T. 1977. The conservation of crop genetic resources: international actions in long term seed storage. Seed Science and Technology 6(4): 1053-1058.

Toll, J. N., Anishetty, M. and Ayad, G. 1980. Directory of germplasm collections: 3: Cereals III: Rice. Rome, FAO, AGR/BPGR/80/109:20 p.

UNEP, 1980. Genetic resources: an overview. Nairobi, UNEP Report No. 5: 132 p.

Van Osten, R. (ed.). 1972. World National Parks: progress and opportunities. Brussels, Hayez. 392 p.

Vera, L. 1980. Tourism and Conservation: The Experience of the World Bank. Brussels, European Architectural Heritage Congress:8 p.

World Bank, 1982. Tribal peoples and economic development: human ecologic considerations. Washington, D.C., World Bank, 111 p.

World Bank, 1983. Incorporation of wildland management in appropriate projects: a policy proposal. Washington, D.C., World Bank, Office of Environmental Affairs (draft ms): 100+p + annexes.

25
Soil Erosion

"Many past civilizations have fallen with their forests and eroded with their soils."   --Edmund G. Brown, Jr.

All too often, modern agriculture "mines" the soil. For all practical purposes, topsoil is a nonrenewable resource like petroleum. In the United States, the average rate of topsoil erosion is about one inch every 30 years. Nature, however, takes 100 to 1000 years to replace an inch of topsoil. In tropical areas lacking winter's protection and exposed to 1-2 meters of warm solvent rain per year, a loss of one inch in two years is not uncommon.

Both water and wind remove enormous quantities of soil each year from agricultural lands. For example, 35 million hectares in the United States lose more than 17 tons of soil per hectare each year (Siddoway and Barnett, 1976); 40 million hectares have already been damaged beyond any practical repair (Constantinesco, 1976). In India, one third of the arable land area is seriously threatened with complete loss of topsoil (Ehrlich, Ehrlich, and Holdren, 1977).

Soil erosion is unusual among economic and environmental concerns because few people argue that the problem does not exist, that it is not serious, or that it needs more study before action is taken. Nor is there any mystery about the physical measures needed to prevent erosion or to stop it once it has begun. An analogy can be made between the spreading, destructive effects of soil erosion and cancer; unlike the case with cancer, however, both the causes and the cures for soil erosion are well known.

The obvious signs of severe water erosion are often dramatic; they include gullies, exposure of bedrock, and sedimentation of rivers and dams. Less obvious signs on farms and grazing and forest lands are decreasing yields and diminishing water supplies (Constantinesco, 1976). The increased rate of rainwater release from denuded catchment areas often leads to catastrophic flooding after heavy rains. Crop production on fertile floodplains thus becomes increasingly risky and further crop losses result.

Factors Influencing Erosion Rates: The extent to which soil is eroded by water depends upon the nature of the soil, the gradient and length of the slope, climate, cultural practices, and (especially) the extent of plant cover. Soils that are especially liable to erosion include clay soils with low infiltration rates, most soils with low organic content, and those soils which lack a well-developed crumb structure. The steeper the gradient and the longer the slope over which surface-flowing water will acquire speed, the more soil will be detached and carried away. The greater the intensity of the rainfall (especially after the soil has been well wet), the more serious the erosion. The timing and nature of cultivation determine how loose and erodible the soil surface is when intense showers and storms occur. Vegetation and plant residues act as a buffer that protects the soil surface from the destructive force of raindrops. Residues also act as a sponge, reducing the volume of surface flow.

The seriousness of wind erosion also depends on the nature of the soil, climate (wind force is especially important when the soil is dry), and cultural practices. The distance over which the wind can move without meeting obstacles is more important than the slope of the land itself. The most wind-erodible soils are usually dry sands, containing a high proportion of loose particles or aggregates small enough to be rolled or carried in the wind. Again, the presence of vegetation and plant residues can protect the soil. Certain cultivation practices are also effective.

"Acceptable" rates of erosion, according to the United States Soil Conservation Service, range from 4.5 to 11.2 tons/ha/year (which seems to be very high), but the acceptability of such figures depends on the nature of the soil and other conditions (Siddoway and Barnett, 1976). If the soil is thin and overlies a very slowly weathered parent material, an acceptable rate would more prudently be near zero. Under agricultural conditions, new soil is typically formed at a rate of 1.1 tons/ha/year. The United States values quoted correspond to an "average" rate of new topsoil formation of about 4 tons/ha/year, according to Carter (1977).

Erosion Control Strategies: A series of measures is available for the control of water and wind erosion (see references). To achieve a stabilized landscape with sustainable productivity, a very flexible approach may be necessary. Because of the varying needs and potentials within a project area, its various sections may best be protected in different ways, by employing a range of methods. Well in advance of any change of land use, the topography, soils, vegetation, and rainfall and drainage patterns in the project area should be napped and present rates and types of erosion should be measured. Estimates of the probabilities of various rain intensities and/or wind speeds are essential for all areas. After project implementation has begun, unexpectedly high sediment loads occur in some streams, or high soil loss rates occur on some surfaces, additional control measures will be needed. However, prevention is cheaper, easier, and more desirable than cure.

Land shaping, revegetation, and other erosion control measures are well known and essential, but not always effectively implemented. Earth moving or forest clearing should be carried out with the lightest equipment practicable to reduce the risk of soil damage, including scraping, compaction, and erosion. The recessed, compacted track marks left by heavy machinery often serve as channels triggering severe erosion. Using lighter, portable equipment or simply slash-and-burn clearing tends to conserve topsoil and protective vegetation and plant residues much more than land clearing by heavy machinery (Sanchez, 1977).

The timing and sequencing of operations during land preparation can be important in erosion control. Where a plantation crop or pasture is to be established after forest removal, a cover crop should be established early in the first wet season after clearing to avoid most erosion problems. If rain is likely to occur before clearing is complete, erosion losses may be smaller if slopes are cleared from the top downward. Measures which control the intensity of cattle grazing and rotate the position of available watering points will prevent some erosion.

Water Erosion Control Using Vegetation Cover and Mulches: The cheapest and most effective methods of erosion control are often biological. For instance, tall, dense secondary forest may limit soil loss to 0.1 ton/ha/year on a slope of 23 percent, and to 1 ton/ha/year on a slope of 65 percent (Roose, 1977). Methods of erosion control appropriate for intensively cultivated areas may need to incorporate physical as well as biological aspects. These should be rationally planned, regularly monitored, and adjusted until performance criteria are met.

When intense rains follow soon after the onset of the rainy season, a month's delay in planting a cover crop can change the rate of soil loss from an acceptable 1 ton/ha to a severe 2,090 tons/ha, while also increasing rainfall loss through runoff from 5 to 20 percent (Roose, 1977). Under such conditions, the speed of establishment is more critical so that instead of a nitrogen-fixing legume, a fast-growing grass species may be preferable. Grasses such as Pennisetum purpureum, Panicum maximum, or Setaria spp. will usually grow faster than legumes, forming a satisfactory cover at least two months before the legumes. Where the need for erosion control is paramount, rhizomatous grasses (e.g., Cynodon aethiopicus) are preferred to tuft plants (e.g., Panicum maximum) because they create a vegetational network of minute dams. In sites less liable to erosion, ground covers of slower-growing legumes such as Stylosanthes, Crotalaria, or Mimosa invisa are effective. Trees alone (without ground cover) are less reliable in erosion control.

Where a range of crops or other vegetation may be grown for erosion control, the ideal will be a type offering quick growth; long duration; a large and well-dispersed leaf area index (to act as a sponge); close-set stems (to act as a comb); smaller, finely-divided leaves (to reduce the size of drops striking the ground); and leaves with hanging apices (so that a large proportion of rain will drip to the ground from leaf tips slowly). Protective crops include sugar cane and small millets; erosion-prone crops include cotton, tobacco, and maize. Supplementary benefits to be sought in growing non-crops include firewood, fodder, mulch, and materials for non-farm enterprises. Vegetation species to be avoided include those with wind-dispersed seeds or fragmenting stolons that could give rise to weed problems in associated crops, and those that tend to harbor pests.

If sufficient crop residues are generated and left in the field, erosion rates may be very low on slopes ranging from 1 to 15 percent (Lal, 1975). Contour cultivation is unnecessary under heavy mulches. As they encourage earthworm populations, mulches may improve rainwater infiltration and cause the soil to be biologically tilled without mechanically disturbing the mulch surface. The costs of mulching vary with conditions but mulching is always labor intensive. In the Ivory Coast, for example, 150 man-days were required to collect 4-10 tons of Guatemala Grass from a field and spread it over a hectare of arable land (Roose, 1977). Mulching is consequently unpopular with many farmers. However, it is often an appropriate solution in labor surplus areas of the tropics.

Activities which remove vegetation, crop residues, and leaf litter require careful control. On arable land, burying or burning crop residues can enhance operational convenience and the control of pests, diseases, and weeds, but these advantages must be weighed against the increased erosion hazard. Similarly, on grazing land, the short-term gain of fodder through overgrazing or burning must be weighed against the long-term loss of fodder resulting from soil erosion and volatilization of nutrients. Wood gathering for direct burning or charcoal production, in conjunction with intense goat grazing, has led to the degradation of immense areas of semi-arid marginal land. Varying the timing and degree of vegetation removal can be used in the erosion control program. Management strategies for crop and grazing land should be adjusted to variations in erodibility of the land and erosiveness of the climate; such strategies should be constantly monitored to ensure effectiveness (Dasmann, et al., 1973).

Forests are exploited most soundly by carefully and selectively felling commercially valuable tree species, so that the lost cover quickly regenerates. If clear-felling is unavoidable, clearance of strips roughly aligned along contours reduces erosion, while also conserving seed trees for forest regeneration. Alignment of heaps or windrows of forest debris along contours also reduces erosion by encouraging early regeneration of suitably oriented strips of vegetation. Similarly, patches cleared by shifting cultivators cause less erosion if they are strips aligned along contours. If the cleared area is to be farmed permanently (rather than allowed to regenerate under shifting cultivation), establishing a protective cover of crops or pasture will reduce the risk of erosion.

To prevent them from causing erosion, roads should be graded to direct runoff into roadside drains which are protected by rhizomatous grasses and stepped as necessary to reduce flow rates.

Water channels passing under roads need sufficiently large
culverts. Although such precautions are well known, they are
insufficiently used, resulting in continued severe erosion.

Water Erosion Control Through Tillage and Land Forming:
Minimum tillage of cropland has been practiced traditionally in
semi-arid areas (where weeds are relatively easy to control) and in
shifting cultivation. By not disturbing the soil surface between
crops, the farmer gains two main advantages. First, soil is
conserved because crop and weed residues protect the soil surface
from rain drops, thus reducing runoff. In one cropping season on a
15 percent slope, with plowing, erosion was 23.6 tons/ha; whereas
without tillage, it was 0.14 ton/ha (Greenland, 1975). Second,
water infiltration is improved by the presence of vegetation
residues, such that at planting, water stored in the soil may be as
much as 37 percent of the previous rainfall, as compared with only
16 percent without residues (Unger and Stewart, 1976). A
beneficial, self-reinforcing cycle is thereby established, in which
increased residue production further promotes water retention.

Plowing, when practiced, should rigorously follow the
land's contours if erosion is to be checked. In West Africa, grain
crops often respond to deep plowing. Due to better water
infiltration, erosion may be low for some weeks, but thereafter the
condition worsens (Roose, 1977). Disturbance of the soil surface
during rainy periods should be discouraged. Ridging along the
contours also significantly reduces erosion after plowing. Storm
water channels can also prevent erosion damage. Details of such
methods are provided by Schwab, et al. (1955), USDA (1969), and
Constantinesco (1976).

Wind Erosion Control Using Vegetation: Vegetation
windbreaks reduce wind erosion by breaking air flow and lowering
wind speed near the ground. Effective protection extends
horizontally to about ten times the height of the windbreak. For
maximum effect, tall vegetation barriers are grown at right angles
to the direction of the prevailing strong wind at a distance apart
of about ten times their height. Not only is the soil kept in place
by this practice, but under dry conditions the sheltered crop plants
suffer less moisture stress and may yield 20 percent more output
(Radke and Hagstrom, 1976). Suitable species for windbreaks
maintain foliage from the ground level and can be row-planted
closely enough to offer a fairly uniform resistance to wind along
the row. Trees provide more permanent shelter, but double rows of
maize or sunflowers are commonly used to protect such annual crops
as soybeans (e.g., 2 rows of maize with 14 rows of soybeans).
Double rows of tall wheatgrass (Agropyron elongatum), spaced 15
meters apart, can almost halve the wind speed at ground level
(Siddoway and Barnett, 1976). Preferred windbreak species are those
which do not provide refuge or serve as alternate hosts for pests
that affect the main crop (Chapter 18). Windbreak plants also can
be selected for other intrinsic values such as providing fruit or
other crops, mulch, fuelwood, or habitat for (non-pest) birds or
other wildlife.

Although windbreaks are an essential erosion control measure in many areas, one tradeoff is the shading which prevents full utilization of land near the rows. To a certain extent, most windbreak species also compete with the main crop for both moisture and nutrients.

Besides windbreaks, a dense and uniform cover of a fine-leaved crop or other vegetation can provide additional protection from wind erosion. If the landscape is a mosaic of protective and non-protective crops, a reduction in the size of the field planted to non-protective crops can reduce wind erosion, even without windbreaks.

Rehabilitation of Degraded Land: The success of any land rehabilitation effort depends upon controlling the influences which originally caused the degradation. Unless effective land use control is provided, expensive project failure is likely. In overgrazed areas, vegetation can sometimes be restored by simply leaving the surface covered with thorny branches. Also, all such measures depend upon effective fire control. Fire danger can be minimized by ensuring that local people gain nothing exceptional from fire. If products from the planted area benefit enough local people, the people are more likely to protect it from fire. Nonetheless, firebreaks should frequently be included as project components. As an example, mulberry bushes in Java have provided both protection and feed for silkworm culture.

Socio-economic Aspects of Erosion Control: Clearly, the level of investment in erosion control will inevitably be a compromise between the ideal and the feasible. Many farmers tend to maximize short-term yields at the expense of longer-term benefits from soil conservation (Carter, 1977). This tendency is particularly marked in areas where farmers lack secure land tenure or access to agricultural support services (particularly soil conservation extension). Indeed, Constantinesco (1976) notes: "If (the farmers) feel insecure, whether as owners, tenants, or otherwise in occupation of the land, they will tend to "mine" that land until it is no longer productive, and then move elsewhere and repeat the process". In Africa, Ruthenberg (1980) observed that mistreatment of the land seemed to increase in proportion to the farmer's commitment to cash cropping. Where it can be shown that effective erosion control is associated with higher yields in the present crop, it will be easier to interest farmers in control measures (Constantinesco, 1976).

Since the level of farmers' dedication to erosion control will immediately affect many aspects of their lifestyle, and often their income, they must be involved at all stages of project development, including planning, research, and execution. Without such involvement, any technological intervention is unlikely to succeed and may even be counterproductive (Saouma, 1979; King and Chandler, 1978). Where wealth is locally measured by the number of cattle owned (and bride prices are paid using the same scale), controlling overgrazing may be particularly difficult; the necessary changes of farmer behavior may only be achieved by changes in natural resource policy adopted by the national government. A recommendation that Turkey should reduce arable land by one-fifth and herds by one-third in order to control erosion had no effect, presumbly because the government had no altenative way of substituting for the subsistence lost by the people affected (Borgstrom, cited by Ehrlich and Ehrlich, 1977).

Even a combination of effective extension services, government support, and lavish provision of finances in no way guarantees success. For example, despite billions of dollars applied, large areas of the United States remain threatened with a repeat of the "dust bowl effect" (Carter, 1977) because of farmers' neglect of erosion control measures. Fifteen tons of topsoil wash out of the mouth of the Mississippi River every second. So far, erosion is winning the race with ease. Despite the political, socio-economic, and administrative complications associated with soil conservation efforts, the threat of soil erosion to the world's future agricultural output is too great to be neglected any longer.

References

Arnold, J.E.M. 1979. A habitat for more than just trees. Ceres 12 (5):32-37.

Carter, L.T. 1977. Soil erosion: the problem persists despite the billions spent on it. Science 196:409-411.

Constantinesco, I. 1976. Soil conservation for developing countries. FAO, Rome, 92 p.

Dasmann, R.F., Milton, J.P. and Freeman, P.H. 1973. Ecological principles for economic development. London, Wiley, 252 p.

Eckholm, E.P. 1976. Losing ground: environmental stress and world food prospects. New York, Norton, 223 p.

Ehrlich, P.R, Ehrlich, A.H. and Holdren, J.P. 1977. Ecoscience: population, resources, environment. San Francisco, Freeman, 1,051 p.

Greenland, D.J. 1975. Bringing the Green Revolution to the shifting cultivator. Science 190 (4217):841-844.

Greenland, D.J. and Lal, R. (eds.) 1977. Soil conservation and management in the humid tropics. New York, Wiley, 283 p.

Halcrow, H.C., Heady, E.O., and Cotner, M.L. (eds.). 1983. Soil conservation policies, institutions, and incentives. Ankeny, Iowa, Soil Conservation Society of America, 330 p.

Harcharik, D.A. and Kunkle, S.H. 1978. Forest plantation for rehabilitating eroded lands. Rome, FAO Conservation Guide 4:83-101.

Hayward, D.M., Wiles, T.L., and Watson, G.A. 1980. Progress in the development of no-tillage systems for maize and soya. Outlook on Agriculture 10(5):255-261.

Holy, M. 1980. Erosion and environment. Oxford, Pergamon, 225p.

King, K.F.S. and Chandler, M.T. 1978. The wasted lands. Nairobi, International Council of Research in Agroforestry, 35 p.

Lal, R. 1975. Soil erosion problems on alfisols in Western Nigeria and their control. Ibadan, International Institute for Tropical Agriculture Monograph I:208 p. (to VI 1981:215-230).

Lal, R. 1981. Soil erosion as a constraint to crop production. Los Banos, Int. Rice Research Inst:405-423.

Morgan, R. 1979. Soil Erosion. London, Longmans, 113 p.

Pereira, H.C. 1973. Land use and water resources. London, Cambridge Univ. Press, 246 p.

Radke, J.K. and Hagstrom, R.T. 1976. Strip intercropping for wind protection (201-222). in Papendick, R.I., Sanchez, P.A. and Triplett, G.B. (eds.) Multiple cropping. Madison, Wisconsin, Amer. Soc. Agron: 378 p.

Roose, E. 1977. Erosion et ruissellement en Afrique de l'Ouest: vingt annees de measures en petites parcelles experimentales. Travaux et documents de l'ORSTOM, Paris, Orstom 78:108 p.

Ruthenberg, H. 1980. Farming systems in the tropics. Oxford, Clarendon Press, 424 p.

Saha, S. K. and Barrow, C. J. (eds.) 1981. River basin planning: theory and practice. New York, Wiley, 357 p.

Sampson, R.N. 1981. Farmland or wasteland: a time to choose. Emmaus, Pennsylvania, Rodale Press, 422 p.

Sanchez, P.A. 1977. Advances in the management of oxisols and ultisols in tropical South America. Tokyo, Japan, Soc. Sci. Soil & Manure: 535-566.

Saouma, E. 1979. To rescue the marginalized. Ceres 12 (5):3-4.

Schwab, G.O., Frevert, R.K., Edminster, T.W. and Barnes, K.W. 1955. Soil and water conservation engineering. New York, Wiley, 683 p.

Siddoway, F.H. and Barnett, A.P. 1976. Water and wind erosion control aspects of multiple cropping (317-335) in Papendick, R.I., Sanchez, P.A., and Triplett, G.B. (eds.), Multiple Cropping. Madison, Wisconsin, Amer. Soc. Agron., 378 p.

UNDP, 1981. Watershed management: environmental operational guideline 1103. NY, UNDP, G3300-1:29 p.

Unger, P.W. and Stewart, B.A. 1976. Land preparation and seedling establishment practices in multiple cropping systems (255-273) in Papendick, R.I., Sanchez, P.A., and Triplett, G.B. (eds.), Multiple Cropping. Madison, Wisconsin, Amer. Soc. Agron., 378 p.

USDA, 1969. Engineering field manual. Washington, D.C., United States Department of Agriculture, Soil Conservation Service, (v.p.):17 ch.

Vietmeyer, N. 1979. A front line against deforestation. Ceres 12 (5):38-41.

Wilken, G.C. 1977. Integrating forest and small-scale farm systems in Middle America. Agro-Ecosystems 3:291-302.

Epilogue: The Future

"We have not inherited the earth from our fathers;
we are borrowing it from our children".

--David R. Brower.

This book outlines various relatively minor options for
environmental improvement that make little fundamental change in the
prevailing conventional style of tropical agriculture. But it is
hoped that these realistic options will be found useful to designers
wanting to reduce any environmental problems.

At the same time, we acknowledge the growing and, to us,
compelling body of opinion that major changes in agricultural
development strategies are essential and overdue. Most countries
now import part of their grain. At least 100-200 million people are
severely malnourished. Thirty million children under 5 years of age
die each year from malnutrition. Efforts to approach basic food
self-sufficiency continue to fail in many tropical nations. By the
close of this century, the world may have to feed as many as two
billion additional people, mostly living in the tropics and largely
on marginal lands (Plucknett and Smith, 1982). To provide a
minimally acceptable food supply for their people by the year 2000,
developing countries will have to double their own food production.
Furthermore, food exports from food surplus nations need·to be
tripled, according to the 1980 Brandt Commission Report.

Fundamental to this problem--and even more important than
agricultural improvements--is population planning, in both developed
and developing countries. Although population growth rates have
slowed dramatically in most industrialized countries (and have even
approached zero in some), population planning in these countries
continues to be important because people in the developed world
consume a disproportionately large share of the Earth's resources.
The product of population and per capita resource use must be less
than or equal to environmental carrying capacity if our planet is to
avoid an agricultural--and societal--collapse. Augmenting global
carrying capacity by the use of energy (embodied in biocides,
fertilizers, and petroleum-based fuels) and by conversion of
marginal lands already brings diminishing returns in agriculture,
and is unsustainable to the extent that nonrenewable resources are
wasted and renewable ones are degraded or destroyed.

The major changes that are needed will be difficult--some will say impossible--to achieve by traditional approaches. Both unconventional and improved conventional approaches are probably necessary. Limits to per capita natural resource consumption; controlling demand from the affluent, rather than just supply (Hardin, 1981); improved natural resource use efficiency, including recycling; and improved energy, water, and nutrient management are fundamental underpinnings for major improvements in agriculture (and other natural resource-intensive sectors). Similarly, there is vast potential in improving inequitable land tenure patterns, which promote poverty, environmental degradation, and inefficient natural resource use. Such major changes as these will be politically difficult and cannot be accomplished overnight. However, they appear inevitable if large-scale disaster is to be avoided. While this book is limited almost entirely to presenting relatively conventional, easily implemented options, many, more controversial changes (outlined in the following literature) will be necessary if humanity is to live in harmony with the natural environment upon which all life depends.

The world faces the transition from cheap but exhaustible resources, like petroleum, to currently expensive but renewable resources, like solar, biomass energy, and hydroelectricity. We, the global community, must learn to live sustainably within the renewable resource base in a qualitatively improving economy. The choice is between careful planning to make the inevitable transition as smoothly as possible, or letting the inevitable dictate the suddenness and extent of the break (Brown, 1981; Daly, 1980).

Our record of global management is poor. We operate the globe by depleting cheap oil fast and producing food too expensively, while permitting hundreds of millions of our community to suffer hunger throughout their meager existence. As stewards of the global commons, responsible for leaving the world a better place (or, at the very least, no worse than we found it), we are failing. We have inherited the right to enjoy this world during our brief tenure and an obligation to pass it on to future generations at constant or increased value. To maximize present value, thereby liquidating our inherited natural capital, rather than sustaining life on the interest from renewable resources, is to limit sharply the options that we leave to successive generations.

- 228 -

References:

FUTURE DIRECTIONS & ECO-AGRICULTURE

Alternative Agriculture News. Monthly newsletter publ. by Institute for Alternative Agriculture, Greenbelt, Maryland.

Altieri, M.A. et al. 1983. Developing sustainable agro-ecosystems. Bio-Science 33 (1):45-49.

Anderson, R.E. 1979. Biological paths to self-reliance. New York, Van Nostrand Reinhold, 367 p.

Arkcoll, D.B. 1979a. The production of food from trees and forests. San Jose, Costa Rica, Int. Symp Forest Sci. ms:4 p.

Arkcoll, D.B. 1979b. Nutrient recycling as an alternative to shifting cultivation. Berlin, Ecodevelopment and Ecofarming Conf., Berlin Sci. Foundation(ms):11 p.

Attfield, R. 1983. The ethics of environmental concern. New York, Columbia Univ. Press, 220 p.

Bishop, J.P. 1980. Agro-forestry systems for the humid tropics east of the Andes. Cali, Colombia, CIAT International Conference on Amazon Land Use and Agricultural Research, 25 p.

Brady, N.C. 1982. Chemistry and world food supplies. Science 218(4575):847-853.

Brandt, W. (ed.) 1980. North-South: A program for survival (Brandt Commission Report). London, Pan Books, 304 p.

Brown, L. R. 1981. Building a sustainable society. New York, Norton, 433 p.

Brown, L.R. and Shaw, P. 1982. Six steps to a sustainable society. Washington, D.C., World Watch Paper 48:63 p.

Buchanan, A. 1982. Food, poverty and power. Nottingham UK., Spokesman Press, 123 p.

Buttell, F.H. 1980. Agriculture, environment and social change (453-488) in Buttell, F.H. (ed.). Rural Sociology in Advanced Societies.

Carter, V.G. and Dale, T. 1976. Topsoil and civilization. Norman OK., Univ. Oklahoma Press, 293 p.

Casco Montoya, R. 1979. Manejo del agua en un ecosistema tropical. Tabasco, Centro de Ecodesarollo, 70 p.

Ciferri, O. 1981. Let them eat algae (Spirulina). New Scientist (Sept.):810-812.

Conway, G.R. 1979. Ecology in agriculture. London, Imperial College, Centre for Environmental Technol. ms:12 p.

Cristiansen, M.N. and Lewis, C.F. (eds.) 1982. Breeding plants for less favorable environments. New York, Wiley, 459 p.

Dahlberg, K.A. 1979. Beyond the Green Revolution. New York, Plenum, 256 p.

Daly, H.E. 1977. Steady-state economics. San Francisco, Freeman, 185 p.

Daly, H.E. (ed.) 1980. Economics, ecology, ethics. San Francisco, Freeman, 372 p.

Eckholm, E.P. 1982. Down to earth: environment and human needs. New York, Norton, 238 p.

Eckholm, E.P. 1979. The dispossessed of the earth: Land reform and sustainable development. Washington, D.C., Worldwatch Inst., 48 p.

FAO, 1978. The place of forests and trees in integrated rural development. Rome, FAO, COFO-78/3 (April): 6 p.

FAO, 1979. The third agricultural revolution. Rome, FAO PPAB/79/30(IP) 19 Oct.: W/N3200: 11 p.

FAO, 1981. Agriculture: Toward 2000. Rome, FAO, 134 p.

Ford, B. 1978. Future food: alternate protein for the year 2000. New York, Morrow, 300 p.

Gliessman, S.R., Garcia-E.R., and Amador-A.M. 1978. Modulo de produccion diversificada, un agroecosystema de produccion sostenida para el tropico calido-humedo de Mexico. Cardenas, Tabasco, Col. Sup:Agric.Trop:19 p.

Gliessman, S.R. and Amador-A.M. 1979. Ecological aspects of production in traditional agroecosystems in the humid lowland tropics of Mexico. Kuala Lumpur, Fifth Int. Symp. Trop. Ecol. ms: 13 p.

Gliessman, S.R. 1979. The use of some tropical legumes in accelerating recovery of productivity of soils in the lowland humid tropics of Mexico. Washington, D.C. National Acad. Sci. Tropical Legumes, 331 p.

Gliessman, S.R. 1979. Some ecological relationships of traditional agroecosystems in the humid tropics of Southern Mexico. Cardenas, Tabasco, Col. Sup. Agric. Trop., ms: 15 p.

Gliessman, S.R., Garcia, E.R., and Amador, A. M. 1981. The
ecological basis for the application of traditional agricultural
technology in the management of tropical agro-ecosystems.
Agro-Ecosystems 7(3): 173-186.

Greenland, D.J. 1975. Bringing the green revolution to the shifting
cultivator. Science 190:841-844.

Hardin, G.J. 1973. Exploring new ethics for survival. New York,
Viking, 273 p.

Hardin, G.J. 1981. Ending the squanderarchy (147-164) in Daly, H.E.
and Umana, A.F. (eds.) Energy, Economics and the Environment.
Boulder, Colorado, Westview Press, 200 p.

Hardin, G.J. and Baden, J. (eds.) 1977. Managing the commons. San
Francisco, Freeman, 294 p.

Hardin, G.J. 1982. Naked emperors: essays of a taboo stalker. San
Francisco, Kaufmann, 300 p.

Harding, J. 1982. Tools for the soft path. San Francisco, Friends
of the Earth, 288 p.

Hodges, R.D. 1978. The case for biological agriculture. Ecologist
Quarterly (Summer):122-143.

Hulse, J.H. 1982. Food science and nutrition: the gulf between
rich and poor. Science 216:1291-1294.

Janzen, D. H. 1973. Tropical agroecosystems. Science 182:1212-1219.

Kiley-Worthington, M. 1981. Ecological agriculture: what it is and
how it works. Agriculture and Environment 6:349-381.

Kock, W. 1983. Principles and technologies of sustainable
agriculture in tropical areas. Washington, D.C., World Bank,
ms: (in prepn.).

Kuc, J. 1982. Induced immunity to plant disease. BioScience
32(11):854-860.

Lockeretz, W. (ed.) 1983. Environmentally sound agriculture, New
York, Praeger, 462 p.

Loucks, O.L. 1977. Emergence of research on agroecosystems. Ann.
Rev. Ecol. Syst. 8:173-192.

Maier, E. 1979. Chinampa tropical. Tabasco, Centro de
Ecodesarrollo, 89 p.

Merrill, R. (ed.) 1976. Radical agriculture. New York, Harper and
Row, 459 p.

NAS, 1979. Microbial processes: promising technologies for developing countries. Washington, D.C., National Academy of Sciences, 198 p.

Orozco-Segovia, A.D.L. and Gliessman, S.R. 1979. The Marceno in flood-prone regions of Tabasco, Mexico. Vancouver, XLXIII Congr. Americanists: 17 p.

Peck, R.B. 1977. Sistemas agro-silvo-pastoriles como una alternativa para la reforestacion en los tropicos Americanos. Bogota, Colombia, CONIF:73-84 p.

Perelman, M. 1977. Farming for profit in a hungry world. New York, Universe Books, 238 p.

Pimentel, D. et al. 1980. Environmental quality and natural biota. Ithaca, N.Y., Cornell University, (Entomol.) Report 80-1:43 p.

Pirie, N.W. 1969. Food resources: conventional and novel. Middlesex, England, Penguin Books, 208 p.

Plucknett, D.L. and Smith, N.J.H. 1982. Agricultural research and third world food production. Science 217:215-220.

Revelle, R. 1976. The resources available for agriculture. Scientific American (Sept.): 164-178.

Rodale, R. 1983. The search for a sustainable agriculture. The Futurist (Feb.):15-20.

Romanini, C. 1976. Ecotecnicas para el tropico humedo. Mexico, D.F., Centro de Ecodesarrollo, 184 p.

Saouma, E. 1979. Promoting the rational use of natural resources. Rome, FAO, W/N 19784/c: 11 p.

Shaner, W.W., Philipp, P.F. and Schmehl, W.R. 1982. Farming systems research and development: guidelines for developing countries. Boulder, Co., 414 p.

Shepard, M. and Jeavons, J. 1977. Appropriate agriculture. Menlo Park, California, Intermediate Technology, 14 p.

Soemarwoto, O. 1975. Rural ecology and development in Java (275-281) in van Dobben, E.H. and Lowe-McConnell, R.H. (eds.) Unifying concepts in ecology. The Hague, Junk, 302 p.

Trenbath, B. R. 1974. Biomass productivity in mixtures. Adv. Agron. 26:177-210.

Tudge, C. 1977. The famine business. London, Faber & Faber, 141 p.

UNEP, 1982. Non-energy uses of agricultural and agro-industrial residues. Paris, Industry and Environment 5 (2):1-34.

UNICEF, 1981. The UNICEF home gardens handbook: for people promoting mixed gardening in the humid tropics. New York, UNICEF:55 p.

USDA, 1980. Report and recommendations on organic farming. Washington, D.C., USDA:94 p.

Wijewardene, R. 1980. 'No till': a tropical farming revolution. Mazingira 4(1):60-63.

Wittwer, S.H. 1979. Agriculture for the 21st century. (Ninth Coromandel Lecture, New Delhi). East Lansing, MI. Agric.Expt.Station, ms:76 p.

Wittwer, S.H. 1980. Agriculture in the 21st century. East Lansing, MI. Agric.Exp. Station, ms: 62 p.

Wyatt-Smith, J. 1979. Agro-forestry in the tropics: a new emphasis in rural development. Span 22:65-67.

About the Authors

Robert Goodland is a tropical ecologist in the Office of Environmental Affairs of the World Bank in Washington, D.C. He started environmental assessments of tropical development projects for the World Bank in 1972, while creating a Department of Ecology in the new University of Brasilia. He has held professorships there, in Manaus, in Costa Rica, and at McGill University in Montreal. His previous books concern environmental aspects of tribal peoples, buildings, power transmission lines, the Trans-Amazon Highway, the cerrado ecosystem of Brazil, and tropical hydroprojects.

Catharine Watson was a researcher in the Office of Environmental Affairs at the World Bank in Washington, D.C. from 1979 to 1980, during which time she assembled much of the material on which this book is based. Since then she has been free-lancing on health issues, and on development policy in the Third World.

George Ledec is a research assistant in the Office of Environmental Affairs of the World Bank in Washington, D.C. His background includes both tropical ecology and economic development studies. He has had field experience related to tropical agriculture in Indonesia, Papua New Guinea, Costa Rica, and Kenya. His previous writings concern tropical deforestation and wildland management.

NOTE: The personal opinions expressed in this book do not necessarily reflect the official position of the World Bank. Neither the facts nor their interpretation should be attributed to the World Bank.

Index